GLOBAL PLANNING INNOVATIONS FOR URBAN SUSTAINABILITY

As the world becomes more urbanised, solutions are required to solve current challenges for the three arenas of sustainability: social sustainability, environmental sustainability and urban economic sustainability.

This edited volume interrogates innovative solutions for sustainability in cities around the world. The book draws on a group of 12 international case studies, including Vancouver and Calgary in Canada, San Francisco and Los Angeles in the US (North America), Yogyakarta in Indonesia, Seoul in Korea (South-East Asia), Medellin in Colombia (South America), Helsinki in Finland, Freiburg in Germany and Seville in Spain (Europe). Each case study provides key facts about the city, presents the particular urban sustainability challenge and the planning innovation process and examines what trade-offs were made between social, environmental and economic sustainability. Importantly, the book analyses to what extent these planning innovations can be translated from one context to another.

This book will be essential reading for students, academics and practitioners of urban planning, urban sustainability, urban geography, architecture, urban design, environmental sciences, urban studies and politics.

Sébastien Darchen is Lecturer in Planning at the School of Earth and Environmental Sciences at the University of Queensland, Australia.

Glen Searle is Honorary Associate Professor at the School of Earth and Environmental Sciences at the University of Queensland, Australia, and at the Faculty of Architecture, Design and Planning at the University of Sydney, Australia.

Routledge Studies in Sustainability

https://www.routledge.com/Routledge-Studies-in-Sustainability/book-series/RSSTY

GLOBAL PLANNING INNOVATIONS FOR URBAN SUSTAINABILITY

Edited by Sébastien Darchen and Glen Searle

Routledge
Taylor & Francis Group

LONDON AND NEW YORK

earthscan
from Routledge

First published 2019
by Routledge
2 Park Square, Milton Park, Abingdon, Oxon OX14 4RN

and by Routledge
52 Vanderbilt Avenue, New York, NY 10017

Routledge is an imprint of the Taylor & Francis Group, an informa business

British Library Cataloguing-in-Publication Data
A catalogue record for this book is available from the British Library

Library of Congress Cataloging-in-Publication Data
A catalog record has been requested for this book

ISBN: 978-0-8153-5756-8 (hbk)
ISBN: 978-0-8153-5757-5 (pbk)
ISBN: 978-1-351-12422-5 (ebk)

Typeset in Bembo
by Swales & Willis Ltd, Exeter, Devon, UK

CONTENTS

ILLUSTRATIONS

Figures

Tables

LIST OF CONTRIBUTORS

Tom Becker

Tom Becker is a research assistant at the Institute of Geography and Spatial Planning at the University of Luxembourg. He holds a Master's degree in Social Sciences with a major in Human Geography from Umeå Universitet (Sweden) and a bachelor's degree in History from Queen Mary, University of London (UK). Currently he is a doctoral candidate at the Department of Geography at Ghent University (Belgium). From 2008 to 2014 he was in charge of the Cellule nationale d'Information pour la Politique Urbaine (CIPU – National Information Unit for Urban Policy) in Luxembourg. His research interests include urban policy and urban planning in Luxembourg and in Europe, urban policy transfer and policy mobilities, evidence-based policy-making as well as integrated and sustainable urban policies.

Yale Belanger

Dr Yale Belanger (PhD) is a Professor of Political Science and Adjunct Associate Professor, Faculty of Health Sciences, University of Lethbridge, where he taught Native American Studies (NAS) for 10 years. His doctoral work at Trent University focused on the emergence and evolution of Aboriginal political organisations in late nineteenth- and early twentieth-century Canada. Widely published in various edited compilations and in journals such as *Canadian Public Policy, Aboriginal Policy Studies, Canadian Journal of Criminology and Criminal Justice, International Journal of Canadian Studies, Canadian Foreign Policy, American Indian Culture and Research Journal, Canadian Journal of Native Studies, Native Studies Review* and *American Indian Quarterly*, Dr Belanger sits as a Regional Advisory Board member with the Alberta Rural Development Network's (ARDN) Homelessness Partnering Strategy, is a member of the Canadian Homelessness Research Network (CHRN) and Canadian Observatory on Homelessness (COH), an editorial board member for the Australia

Housing and Urban Research Institute (AHURI) and is a former member of the Alberta Homelessness Research Consortium (AHRC).

Anna Broberg

Anna is a planning geographer by origin. As part of her PhD studies on urban structures and children's mobility at Aalto University, she took part in developing a web mapping survey tool (softGIS) to aid public participation, which later evolved to Maptionnaire. In her academic career she focused on developing new ways of analysing PPGIS data together with registered-based data statistically and geographically. Currently, she strives to help people and organisations around the world to co-create better urban milieus with the help of digital mapping technology.

Manuel Calvo-Salazar

Manuel is a socio-ecologist, he holds a degree in Biology and specialised in Applied Ecology. He holds a Master's degree in Humanities. Manuel works as a consultant on sustainability in the company EstudioMC. His work is focused on the development and implementation of technical assistance in many different fields related to sustainability: sustainable mobility, ecourbanism, climate change, ecological footprint calculations and the introduction of sustainability criteria in strategic planning and strategic environmental assessments. He is author of several publications on ecological footprint, sustainable mobility and urban sustainability. He is also an active speaker on topics related to sustainable mobility, climate change, sustainability, ecological footprint, energy development and urban planning, among others.

Ji-in Chang

Ji-in Chang is Assistant Professor at the Graduate School of Smart City Science Management, Hongik University, where she has taught courses on smart cities and environmentally friendly housing. Chang holds degrees from Seoul National University (PhD in Urban Planning), Massachusetts Institute of Technology (MSc in Architecture Studies) and the University of New South Wales (BArch., Honors). Before joining Hongik University, she was Visiting Research Fellow and one of the founding members of the Megacity Research Center at Seoul Institute, the think tank of the Seoul Metropolitan Government. Chang's research focuses on international connections and the built environment. Topics of particular interest include smart cities, sustainable development, transnationalism and gendered urbanism. She has published in Korean and international journals, and co-translated Linda McDowell's *Gender, Identity and Place* from English to Korean.

Tathagata Chatterji

Dr Chatterji is Professor of Urban Management and Governance at Xavier University, Bhubaneswar. He graduated in Architecture from Bengal Engineering College, Shibpur, Calcutta University, did post-graduation in Urban Design from Kent State University, USA and a doctorate in Urban Planning and Governance, from the University of Queensland Australia. He is a Member of the Planning Institute of Australia. Dr Chatterji had written extensively on contemporary urban issues and challenges. He has published two books, *Local Mediation of Global Forces in Transformation of the Urban Fringe* and *Citadels of Glass: India's New Suburban Landscape* and several peer-reviewed journal articles and book chapters. He also writes op-ed articles on urban issues for reputed national newspapers. He received the prestigious Gerd Albers Award from the International Society of City and Regional Planners (ISOCARP) in 2016 for his research publication on comparative modes of urban governance in India.

Sébastien Darchen

Sébastien Darchen is Lecturer in Planning at the School of Earth and Environmental Sciences (SEES), University of Queensland. He holds a PhD in Urban Studies obtained in 2008 from INRS (Urbanisation, Culture et Société) (Montréal, Canada). Dr. Darchen studies the political economy of the built environment with a focus on the strategies of the urban stakeholders involved in the provision of the built environment (Developers, City Planners, Urban Designers, etc.). His research focuses on the regeneration of inner city areas in France, Canada and Australia and helps improve the sustainability of cities.

Katherine A. Dekruyf

Katherine Dekruyf graduated with a Bachelor of Arts majoring in Native American Studies (now Indigenous Studies) from the University of Lethbridge in 2014. She went on to pursue a Master of Arts degree from the University of Lethbridge specialising in urban Indigenous community development and municipal policy, obtained in 2017. She is now further pursuing her interests in urban Indigenous policy issues as a Communications and Community Liaison with the City of Calgary.

Carl Grodach

Carl Grodach is Professor and Director of Urban Planning and Design at Monash University. His research focuses on the urban development impacts of the arts and cultural economy. He is co-editor of *The Politics of Urban Cultural Policy: Global Perspectives* (Routledge, 2013) and co-author of *Urban Revitalization: Remaking Cities in a Changing World* (Routledge, 2016). His research has been funded by the Australian Research Council, National Endowment for the Arts, New York Community Trust and the Government of Canada.

Peter V. Hall

Peter V. Hall is Professor and Director of the Urban Studies Program at SFU. His research seeks to understand the role of institutions and institutional change in shaping patterns and outcomes of economic development. His wide-ranging areas of exploration include seaports, logistics, social enterprise and community development.

Markus Hesse

Markus Hesse is Professor of Urban Studies at the University of Luxembourg. With an academic background in human geography and spatial planning, his research interests focus on related topics in urban and metropolitan regards. One thread of his research is concerned with the interplay of spaces and flows, for example in a recent research project that studies the emergence of 'relational cities' in Europe and South-East Asia. Also, particular emphasis has been placed on the science-policy interface in urban planning and governance. Markus Hesse is an elected member of the Academy for Spatial Research and Planning (ARL) in Germany and sits on various advisory boards and scientific councils. He is co-editor of *Cities, Regions and Flows* (Routledge, 2013).

Iswanto

Dr Iswanto is an Environmental Health Lecturer in Polytechnic of Health, Yogyakarta since 1999. He is a board member of the Indonesia Environmental Health Specialist Association (HAKLI), and a pioneer of community-based solid waste management in Sukunan Village and Indonesia. He was the head of Community-Based Solid Waste Management Association in Yogyakarta Province (2009–2015). He has received several awards from the Indonesian Ministry of Environment and the Indonesia Ministry of Women Empowerment (2004) and the best lecturer at the Polytechnic of Health Yogyakarta (2004).

Maarit Kahila-Tani

Dr Maarit Kahila is the co-founder of Mapita. Maarit got her PhD in January 2016 in urban planning and public participation. In her thesis she studied the ways in which PPGIS tools and experiential knowledge gathered from people can support urban planning projects and processes. Currently she is contributing to academia by co-writing journal articles with colleagues as well as working as a co-researcher in various research projects. At Mapita she is motivated to learn how urban planners around the globe use Maptionnaire in communicating with local people.

Marketta Kyttä

Marketta Kyttä works as an Associate Professor in the Land Use Planning and Urban Studies Group (YTK) of Aalto University, Finland. Her research interests within environmental psychology and participatory planning cover various topics: child- and human-friendly environments, environments that promote wellbeing and health, and new methods for participation. Currently, her multidisciplinary research team concentrates on the place-based person-environment research with methodological development work. The so-called softGIS studies enable the provision of people's place-based perceptual knowledge in the GIS format, as complementary to the traditional GIS-based types of environmental and demo-graphic data. The Internet-based softGIS methodology developed by the softGIS

team enables the combination of 'soft' subjective data to be produced by inform-
ants with 'hard' and objective GIS-data. Geospatial World Forum granted softGIS
team a webGIS Innovation award in 2010. The softGIS team collaborates with
various actors in the field of urban planning, mainly with cities and governmental and non-governmental organisations but also with private companies. Recent
research themes include the localised study of perceived environmental quality, social
sustainability and the health promoting qualities of urban structure. This research is
associated with international research developments under the headings of Public
Participation GIS (PPGIS) and Post-occupancy Evaluation (POE).

Annick Leick

Annick Leick is currently working as a FNS senior researcher at the University of
Lausanne. She holds a PhD in Geography from University of Luxembourg. Dr.
Leick's work, broadly speaking, focuses on spatial planning and urban development
policies and the related decision-making, planning and implementation processes.
More precisely she examines governance structures and practices as well as discourses and rationales reproduced for legitimating and evaluating spatial and urban
planning policies. She is currently involved in the project Mega-events as urban
interventions: growth and impact that aims to identify patterns and trends in the
growth and impacts of mega-events, to find possible explanations for the presence
or absence, magnitude and nature of impacts and to examine the ways in which
stakeholders, observers and critics have addressed the issues of size and impacts in
different places at different points in time.

Saugata Maitra

Saugata Maitra is National Director – Infrastructure Services at Jones Lang
LasSalle, Kolkata, India. He is responsible for managing and developing business
of the Government and Infrastructure Advisory Practice for the firm in India and
neighbouring markets, and leading public private partnership transaction advisory
engagements in the region. He has over 22 years of experience in project management
and public private partnerships in housing, livelihood, poverty reduction and urban
infrastructure sectors. He holds a Post-graduate Diploma in Business Administration
(PGDBA), ICFAI Business School, Kolkata, India; Post-graduate Diploma in
Business Management (PGDBM) Indian Institute of Social Welfare and Business
Management (IISWBM), Kolkata, India; Diploma in Business Finance (DBF),
Institute of Chartered Financial Analysts of India (ICFAI), India; and a Bachelor of
Engineering (Civil), Manipal Institute of Technology (Mangalore University), India.

Ricardo Marqués

Ricardo Marqués is Full Professor of Physics at the University of Seville, where he
has developed most of his teaching and research activities, authoring more than a
hundred scientific papers and numerous contributions to scientific meetings. He
has been also member of the Board of ConBici – the main Spanish association of
urban cyclists – and has been actively involved in the design of the cycling policy at

his university and city (Seville). Presently, his scientific interest focuses on the bicycle as a sustainable mode of transport, being the author of several reports, scientific articles and contributions to conferences analysing utilitarian cycling in metropolitan areas and cities. His present areas of interest are intermodality between bicycles and public transport, cycling safety and the analysis of the factors that contribute to the integration of cycling into urban mobility.

Declan Martin

Declan Martin is a PhD student in Urban Planning and Design at Monash University. His work investigates the impact of urban development on cultural production, currently focusing on the interaction between material cultural product industries and contemporary modes of urban manufacturing.

Amanda Napoli

Amanda Napoli currently works in transportation planning at the Ontario Ministry of Transportation. She holds a Master in Environmental Studies, Urban and Regional Planning (York University, Toronto) and has worked with a variety of public and private entities around the Greater Toronto Area in areas of planning, policy and public consultation. Her present work is in the development of transportation planning policy and area-wide transportation plans and strategies. Previously, Amanda worked on the implementation of the tolling and back office services for a network of new provincially owned highways. She supported ministries and agencies in program and project delivery at the Ontario Ministry of Infrastructure. Amanda has worked on an internationally recognised public consultation team on consultation efforts as part of a comprehensive master plan update in Toronto. Amanda's previous research focused on the application of alternative zoning codes and retrofitting tactics in North American suburbs, with a particular focus on the Greater Toronto Area.

Sita Rahmani

Sita Rahmani has a Bachelor of Engineering in Urban and Regional Planning degree from Gadjah Mada University (UGM), Indonesia. Her main interest is in environmental management and regional development issues. She has worked as research assistant in several projects, including projects funded by the University of Queensland, Australia. Sita is now affiliated to the Department of Transdisciplinary Science and Engineering at the Tokyo Institute of Technology in Japan.

Wesley Regan

Wesley Regan is Community Economic Development Planner for the City of Vancouver, is the Green Party of Canada's critic for urban affairs and housing, and a Graduate Student from Simon Fraser University's Urban Studies Program. His thesis research examined the impact of mixed-use real estate development on the retail culture of Vancouver's Main Street, a major neighbourhood-serving commercial corridor.

Sun-Young Rieh

Sun-Young Rieh has been a Professor in the Department of Architecture, College of Urban Sciences, University of Seoul since 1998. She is also a registered architect in both South Korea and the US. Rieh studied architecture in Seoul National University and earned a Master in Architecture degree from the University of California, Berkeley. She received a Arch.D degree from University of Hawaii. In 2007, she was the Fulbright Visiting Scholar at the School of Architecture, University of Hawaii. In 2017, she was a Visiting Researcher at the Faculty of Architecture and the Built Environment, TU Delft. Her research interests include sustainable environment, gendered urbanism and educational facilities. She is on the review panel for G-SEED (Green Standard for Energy and Environmental Design). She served as a board member for Architectural Institute of Korea (AIK), Korean Institute of Architects (KIA) and Korea Green Building Council (KGBC). She was Vice President of Korea Institute of Ecological Architecture and Environment (KIEAE).

Sonia Roitman

Dr. Roitman joined the University of Queensland in 2013 following previous academic, research and professional appointments at University College London (UK), School of African and Oriental Studies (SOAS, UK), Free University Berlin (Germany), Universidad Nacional de Cuyo (Argentina), Consejo Nacional de Ciencia y Técnica (CONICET-Argentina), and Secretaría de Ambiente y Desarrollo Sustentable, Gobierno de Mendoza (Argentina). She has lived and worked in Argentina, United Kingdom, Germany, Mexico and Uganda. Her contributions to the field of development planning and urban sociology include influential research on housing policies; gated communities; urban segregation and social exclusion; urban planning, edge cities and growth; international development and poverty alleviation policies.

Thorsten Schuetze

Thorsten Schuetze received his engineering doctorate from the Architecture and Landscape Department of the Leibniz University Hannover in 2005. Since 2012 he is Professor for Architectural Design, at the Department of Architecture of the SungKyunKwan University (SKKU) in Korea. He is also a registered freelancing architect, consultant, researcher and lecturer in the field of sustainable architecture and urbanism. Professor Schuetze coordinated international research projects, is a member of scientific committees and editorial boards and is the author and editor of numerous international scientific publications. Thorsten Schuetze is an active member of the International Forum on Urbanism (IFoU) since he was Assistant Professor at the Faculty of Architecture of Delft University of Technology in The Netherlands from 2006 to 2012.

Glen Searle

Glen Searle is Honorary Associate Professor of Planning at the Universities of Sydney and Queensland. He was formerly Director of the Planning Programs at the

University of Technology, Sydney and the University of Queensland. He has held urban policy and planning positions at the UK Department of the Environment and at the New South Wales departments of Decentralisation and Development, Treasury, and Planning, where he was Manager, Policy.

Luisa Sotomayor

Dr. Luisa Sotomayor is an Assistant Lecturer and the Coordinator of the Graduate Program in Planning at York University. She holds a PhD and a MSc in Planning from the University of Toronto, and a BA in Sociology from the National University of Colombia. Her broad areas of teaching and research include urban policy, affordable housing, city governance and Latin American urbanism. Before starting her PhD program, she worked for several institutions in both Canada and Colombia on issues related to planning, poverty reduction, youth unemployment, and community development.

Donald R. Spivack

Donald R. Spivack is an Adjunct Professor in the Sol Price School of Social Policy at the University of Southern California, where he teaches community and economic development classes at both the undergraduate and graduate level. Mr Spivack holds a Bachelor of Arts in Architecture from the University of Pennsylvania and a Master's of City Planning from Yale University. He is a charter member of the American Institute of Certified Planners, a life member of the Institute of Transportation Engineers, and a Fellow of the Royal Society for the Encouragement of Arts, Manufacture and Commerce. In addition to transportation system development and planning work in Boston, Detroit, Philadelphia and Washington, DC, Mr Spivack served on staff of the Community Redevelopment Agency of the City of Los Angeles (CRA/LA) for nearly 30 years, retiring in 2010 as Deputy Chief of Operations and Policy. In that capacity he oversaw the formation and implementation of long-range redevelopment and revitalisation policies and strategies, including land use, industrial development, transportation and transit-oriented development, job development, business attraction, housing, open space and financing options in an era of declining tax revenues.

Ryan Walker

Ryan Walker is Professor in the Department of Geography and Planning, Chair of the Regional and Urban Planning Program, and has acted as Associate Dean of the College of Graduate and Postdoctoral Studies at the University of Saskatchewan, Canada. A registered professional planner, Walker holds degrees from Queen's University (PhD), University of Waterloo (MA) and University of Lethbridge (BA). Author of roughly 50 publications, he also serves on the Plan Canada editorial board and is co-editor and co-author of the books *Canadian Cities in Transition* (2010, 2015, forthcoming 2020) and *Reclaiming Indigenous Planning* (2013).

1

PLANNING INNOVATION POLITICS AND PROCESS FOR URBAN SUSTAINABILITY

Glen Searle and Sébastien Darchen

Overarching aim

As the world becomes more and more urbanised, it is urgent to propose solutions to solve current sustainability challenges for the three arenas of sustainability: social sustainability, environmental sustainability and urban economic sustainability. This edited volume focuses on the *politics of sustainability* by examining the socio-political, economic and institutional context of urban sustainability innovations and how this shapes the trade-offs made between the three arenas of sustainability in the implementation of innovative solutions. As the book analyses the procedural side of sustainability it offers a new perspective on the implementation of sustainability that will be useful and attractive to both planning professionals and academics. The analysis of sustainable solutions to current challenges is centred on the concept of innovation. Innovative solutions are part of a sustainability transition process. This process is characterised by contextual factors that are city-specific. But this process, in a globalised world, can also be influenced by external factors (such as urban policies or planning models applied in other contexts) and/or the networks in which a city might be involved that facilitate the circulation of planning solutions from one context to another. The book's structure is based on the recognition that planning solutions for sustainability cross three arenas: the social arena, the environmental arena and the urban economic arena. Finally, the book analyses to what extent planning innovations can be transferred from one context to another: we thus reference the *transferability* of innovations.

Defining urban sustainability

Why is yet another book on urban sustainability needed? Thirty years after the Brundtland report, planners still focus on sustainability. Sustainability is considered as a resilient, sustainable idea (Campbell 2016, 392). Hirt (2016, 383) states that the "City Sustainable" – defined as a paradigm of ideas and practices centred on the notion of sustainability – has become a dominant school of thought since the 1990s. However, sustainable development has not necessarily emerged as a dominant planning paradigm for most cities (Saha and Paterson 2008) and in many cases it is still unclear how cities incorporate this paradigm in urban policies.

Therefore, an important question remains: *Has the concept of sustainability made a difference in urban planning practice?*

While there are many possible responses to the question, this book addresses it by examining the procedural side of sustainability. In this, it encompasses all three dimensions of sustainable development (equity/social justice, economic development, environmental protection). We use the term *sustainable planning* to acknowledge that a planning solution implies trade-offs between the three arenas of sustainability. In a recent article, Campbell (2016) argues the need to reassess the triangle of sustainability by recognising the development of conflicts in achieving planning priorities (between social justice and environmental protection for example). We refer to the term trade-offs to illustrate the tension between several core motivations of urban planning. It also reflects Jane Jacob's notion of a sustainable precinct where the solution to one sustainability challenge can have an impact on other problems (Meijer and Thaens 2018).

We use the definition of sustainability developed by Scott Campbell because his dissection of sustainability makes the concept more operational for the planning profession (Campbell 1996; Hirt 2016, 383). We agree with his notion that 'sustainability' is a dynamic concept, unpredictable and plagued with contradictions and ever evolving (Campbell 2016, 396). Thus, Campbell (2016) defines 'sustainability' as the never-ending process of resolving three conflicts over the three planning priorities (equity and social justice; economic development; and environmental protection) to be achieved.

Based on the conflicts identified by Campbell (1996, 2016), this book aims to answer the following sub-research questions on *trade-offs* during the implementation process of sustainability:

- *How are trade-offs between the economic, social and environmental dimensions of sustainable urban development being managed in planning practice?*
- *What types of mechanisms/factors facilitate the management of conflicts related to decisions on trade-offs?*

Our book contributes to the understanding of factors facilitating the implementation of sustainable solutions in planning practice. Compared to most recent works on sustainable development (such as Atkinson et al. 2014; Baker 2016) focusing on the theory of urban sustainability, this book offers a practical understanding of the implementation of planning solutions. While some recent textbooks have focused on solutions (such as Roseland 2012), they have not systematically contextualised those solutions for a given political and institutional situation.

Often, edited volumes on urban sustainability comprise a set of international case studies without a clear framework that connects the different cities together (e.g. Vojnovic 2012). Some good books have focused on sustainability in urban design and architecture (Bovill 2014) but as yet there is not an edited volume presenting and analysing, in terms of replicability and trade-offs, planning solutions that have had a significant impact on current urban sustainability challenges. This book builds on previous pioneering works studying sustainability and its relationship to cities (Newman and Kenworthy 1999). The meaning for planning of each of the three sustainability arenas – equity, economy and environment – is presented in the following sections.

Urban sustainability is an evolving concept and the book provides planning responses for 8 of the 17 sustainability development goals presented in the recent 2030 Agenda for Sustainable Development published in 2015. This new agenda is guided by the principles of the Charter of the United Nations, and is grounded in the Universal Declaration of Human Rights (United Nations 2015, 10). The sustainability goals from the Charter that are addressed in this book are as follows: Goal 3 Ensure healthy lives and promote well-being for all ages; Goal 7 Ensure access to affordable, reliable, sustainable energy for all; Goal 8 Promote sustained, inclusive and sustainable economic growth; Goal 9 Build resilient infrastructure, promote inclusive and sustainable industrialization and foster innovation; Goal 10 Reduce inequality within and among countries; Goal 11 Make cities inclusive safe and sustainable; Goal 13 Take urgent action to combat climate change and its impacts; and Goal 16 Promote peaceful and inclusive societies for sustainable development.

Social sustainability

In Campbell's latest (2016) version of the triangle, the social side of sustainability is summarised by the words equity and social justice. The latter is also associated with this dimension of sustainability through the concept of 'just cities'. In Campbell's (1996) triangle, the words associated with this priority were social justice, economic opportunity and income equality.

Colantonio and Dixon (2010) have suggested key criteria to delineate the concept of the 'socially sustainable community' are as follows:

- The strengths of social networks;
- The participation in collective community activities;
- Participation of ethnic minorities in planning processes;
- Pride or sense of place;
- Residential stability (versus turnover);
- Security (lack of crime and disorder).

Similar criteria have been listed in other works, such as sense of place, empowerment and participation (Newman and Jennings 2008). In this book, we have included case studies in the social sustainability section that are close to the model of 'community planning': the concept suggests that community planning can produce a more socially sustainable end product (Chapters 5 and 6). Our chapters address most of the themes associated with the social side of sustainability or 'social sustainability':

- Ethnic minority and planning/housing (Chapter 2);
- Real-life participatory urban planning and new technologies (Chapter 3);
- Reduction of urban crime through social urbanism (Chapter 4)
- Equitable sustainability planning in a post-industrial context (Chapter 5).

Environmental sustainability

Originally the term 'environmentally responsible development' was used (World Bank 1992), then the concept of 'environmental sustainability' was developed (Goodland 1995). Goodland (1995, 3) defines it as follows: "it seeks to improve human welfare by protecting the sources of raw materials used for human needs." If we translate this in planning terms, environmental sustainability can be associated with the following themes:

- Creation of ecological mixed-used districts (Chapter 6);
- Management of solid waste (Chapter 7);
- Promotion of active transport (Chapter 8);
- Transformation and re-use of obsolete infrastructures (Chapter 9);
- Regeneration and adaptive re-use of buildings (Chapter 10).

Urban economic sustainability

Urban economic sustainability can be regarded as a city's ability to sustain and reproduce its level of economic development or economic prosperity over time (Campbell 2016, 390). As with the other two dimensions of sustainability, this cannot disregard the other sustainability facets. We cannot "submissively serve the economic mandates of the elite urban growth machine" (Campbell 2016, 396) without simultaneously sustaining the urban workforce (human capital) and environmental systems (natural capital) along with the local economy (financial capital) (Campbell 2016, 394). Nevertheless, achieving this in reality can be challenging. In practice, economic interests usually prevail over environmental concerns, and the latter in turn usually prevail over social justice (Campbell 2016). While communicative planning practices can potentially mediate between communities and development interests via innovative new sustainable practices, the reality can be that they merely allow the wearing down and 'managing' of local communities so that developers can still build what they want (Schweitzer 2016, 378). Thus the challenge for city sustainability is to find innovative ways of planning for economic development that do not reduce, and preferably increase, social and environmental sustainability at the same time.

Our chapters cover significant current themes of urban economic sustainability:

- Implementation of the Smart City strategy (Chapter 11);
- Creation of a new model of sustainable land use (Chapter 12);
- Development of a science hub through brownfield development (Chapter 13).

Defining planning innovation

Our definition of planning innovation is organised around three main components: 1) the context from which urban innovation emerges; 2) the external factors influencing the development of innovation; and 3) the management of uncertainty.

Planning innovation as a process in a context of uncertainty

The book's objective is to present urban practices that are innovative and go beyond 'sustainability fix' practices. The sustainability fix – drawing on Harvey's (1982) concept of territorial fixes – can be thought as a spatially and historically contingent organisation of political interests that allows economic growth and development to continue in the face of social and environmental concerns (Tenemos and McCann 2012). In that sense, sustainability is also a very political concept:

> In its many and divergent guises is a goal, concept, or ideology that can be turned to numerous persuasive purposes and can, therefore, play a crucial role in solving, deferring, or redirecting conflicts between local environmental activist groups and business interests.
>
> *(Tenemos and McCann 2012, 1392)*

This book is based on the assumption that sustainable urban development *is* and *reflects* a political process. Sustainable urban development is "neither neutral nor objective, but – as politics always is – subject to societal negotiations reflecting and producing social conflicts, class differences, inequalities and moments of exclusion" (Mössner 2016, 973). Planning solutions are therefore embedded within a specific institutional, political, cultural and social urban context. Models of sustainable urban development are not fixed (Mössner 2016) and most of all, models of sustainable urban development cannot be replicated from one context to another without negotiations or adaptations. In that sense, planning solutions to overcome sustainability challenges are context-specific.

Urban innovation can be understood as innovative practices within urban environments with the aim of improving those environments (Meijer and Thaens 2016). Innovations are embedded in a specific cultural, political and institutional context. Glaeser (2011) speaks of self-protecting innovations when cities are able to generate the solutions they need to solve their own problems. This edited book presents genuine planning innovations that have 'worked' in a given city context and therefore contribute to advance knowledge on sustainable planning. At present, we know very little about how urban innovations are made. As stated by Bai et al. (2010, 313): "Despite that there are well documented best practices in urban sustainability, very little is known about how they are produced and used." Our book focuses on how planning innovations are produced and then implemented in a given context.

Associated with the development of urban innovation to improve sustainability in a given context, is the issue of the management of uncertainty that is endemic in transitioning to new sustainability practices. Geels (2011) uses the term 'sustainability transitions': "A transition is referred as a long-term socio-technological change on the way societal functions are being fulfilled". Sustainability transitions involve "alterations in the overall configuration of transport, energy, and agri-food systems, which entail technology, policy, markets, consumer practices, infrastructure, cultural meaning and scientific knowledge" (Geels 2011, 24).

Transitions have a 'systemic character', which means that governance arrangements are central in 'managing transitions.' In this regard, Meijer and Hekkert (2007 90) state that

> steering sustainable development is problematic due to the ambivalence of goals, the uncertainty of knowledge about systems dynamics and the distributed power to shape system development . . . the systemic character of transitions implies that a wide diversity of actors is involved and that none of the actors can achieve the transition alone.

Hunt and De Laurentis (2015) highlight the challenge of 'governing change' in achieving 'sustainability transitions' in a context of uncertainty. Thus, this book also analyses how uncertainty related to the implementation of planning innovation is being managed in a given context. It considers the following sub-research question on the management of uncertainty: *How has uncertainty related to the planning innovation been managed?*

Transitions to sustainability: Identification of contextual factors

In this book, we consider sustainable planning innovations as part of a sustainability transition process. Markard et al. (2012, 956) define sustainability transitions as "long-term, multi-dimensional and fundamental transformation processes through which established socio-technical systems shift to more sustainable modes of production and consumption." Socio-technical transitions refer to systemic changes involving a wide range of actors: policy makers, consumers, civil society, engineers, researchers and politicians (Geels 2011, 24). It is essential to note that sustainability transitions – in opposition to many historical transitions – are goal-oriented or purposive. Private actors have limited incentives to address sustainability transitions as the goal ('sustainability') is related to the collective good (Geels 2011, 25). Typically, sustainable transitions rely on the role of public authorities and civil society. They will internalise negative externalities and contribute to change economic frame conditions (Geels 2011, 25).

Most sustainability solutions do not offer obvious user benefits and require changes in policies because vested interests will try to resist such changes. Taking the example of environmental innovations, Geels (2011) demonstrates that innovation will only be able to replace existing systems if incentives such as tax subsidies or changes in regulatory frameworks are implemented. Sustainability, as an ambiguous and contested concept, will trigger debate and negotiation on the orientation of the sustainability transition. It will imply trade-offs.

The literature on transition management explains that every transition project is unique (in terms of context and participants). Loorbach and Rotmans (2010) explain that multiple networks of actors are developed around the transition agenda. Frontrunners, defined as go-getters who initiate the transition process, need support and space for their innovation activities: the concept of 'minimally regulated space' as an experimental zone is thus essential to the transition process. Furthermore, frontrunners need to be empowered (Loorbach and Rotmans 2010, 244). Breakthrough projects led by frontrunners are often used as a starting point for the longer-term sustainability vision. At the same time, the literature emphasises that the transition process is a contested one and faces resistance generated by institutional dynamics and relations of power and politics that can be both local and external (Geels 2014; Meadowcroft 2009; Rutherford and Coutard 2014).

Consequently, the literature on sustainability transitions stresses the centrality of context in determining the evolution of niche innovations (Loorbach and Rotmans 2010; Rutherford and Coutard 2014). Thus contextual factors that shape why particular options are chosen, promote helpful linkages and enable momentum to build need to be identified. The literature enables us to identity key contextual factors to consider in the analysis of the emergence and implementation of planning innovations in a given city context. In this regard, Moore et al. (2018) have used, in introductory fashion, Australian city examples and transitions perspectives to understand change towards sustainability. This book significantly extends that approach by using global examples to understand the specific factors and trade-offs involved in innovative urban sustainability transitions.

The key city context factors in sustainability transitions can be identified by answering questions such as the following:

- *Were incentives (tax subsidies, changes in regulatory frameworks) put in place to enable innovation?*
- *Who were the frontrunners? Were there mechanisms put in place to empower the frontrunners?*
- *What kind of actors' networks developed around the transition agenda?*
- *More generally, what elements of the city's political ecology and economy, in terms of power relations, institutional settings and so on, assisted or were in conflict with the transition agenda and, in particular, the sustainability innovation?*

One lens through which to consider such contextual elements relates to the availability of resources. Successful innovative sustainability practice requires adequate implementation resources. Finance is always a central consideration. In the sustainability domain, calculations are clouded by the failure of the market to properly value externality benefits from sustainability projects, such as reductions in greenhouse gases. This means that public subsidies can be warranted to help make sustainable projects happen. These can take the form of in-kind subsidies, such as free public land, or grants and tax concessions. But the difficulties of assigning money values to positive externalities resulting from sustainability projects also mean that it is usually difficult to determine whether too much finance has been invested in a project and thus whether the project has been 'successful' in terms of the finances involved. Adequate resourcing also includes a suitable governance framework. This includes appropriate legislation and development controls, and an appropriate institutional support structure. Special controls or new implementation agencies might be required in many cases to enable sustainability projects to get off the ground. The case studies in this book therefore make explicit the extent to which suitable finance and governance resources were found or made available in order for those projects to proceed and become successful.

More generally, the availability of such resources in turn reflects the particular institutional and political city context (Savitch and Kantor 2002). City governments with more progressive constituencies and more funding available for sustainability initiatives provide environments that enhance the likelihood of such new initiatives (Mössner 2016). Institutions that are less autonomous and more connected within wider city governance structures are more likely to be open to new sustainability ideas and potential solutions. The economic environment is also potentially critical. Cities with more prosperous and/or growing economies are more able to incorporate new sustainability initiatives as part of infrastructure expansion or because they can be more easily subsidised from the public purse as pilot projects. The role of agency is likewise potentially vital. Factors here include agents/actors to dream up, or identify from elsewhere, innovative sustainability approaches. The role of individual leaders within institutions or community groups to champion new ways of achieving sustainability is very often crucial (c.f. Sotarauta et al. 2017). Individual politicians can also play vital roles in making new practices happen.

External factors and transferability of planning innovation

With the imperative towards fast policy (Peck and Theodore 2015), local policy professionals prioritise sustainability policies that: "seem easily accessible in packaged, readily consumable, and mobile form" (Tenemos and McCann 2012, 1393). A number of works on sustainability in the last decade or so have studied how and why 'local sustainability' fixes are constructed and legitimated (While et al. 2004). This perspective can also be seen as encompassed within a sustainability transitions framework. Recent works in the broader field of urban policies have emphasised the emergence of *mobile urbanism* to describe the transfer of policies from place to place with lightning speed (McCann and Ward 2008). External factors influencing the construction of urban innovations in a specific context might rely on the national and international networks in which cities are involved as emphasised by recent works on the urbanisation process in contemporary society (Darchen 2016; Payre 2010). Associated with this question is the issue of transferability. As stated by Bai et al. (2010), transferability and up-scaling of sustainable practices are difficult to achieve, and ready-made solutions for urban sustainability cannot be applied into other contexts without negotiations and modifications. Again, such considerations are embodied within a sustainability transitions approach. In general, more work is needed to understand how innovative sustainability practices can be transferred. Nevertheless there is now an emerging body of work (Ooi 2008) looking at the transferability of innovative planning practices from one geographic context to another. Our book contributes to expanding knowledge in that specific field of urban research.

Overall this book sets out to answer the following research questions about planning innovations for urban sustainability:

– *Is innovation for sustainable development solely reliant on local planning factors (local institutional factors; local planning instruments; local politics, leaders and community groups; local built and natural environment; economic potential for redevelopment, etc.) or also on other factors not necessarily bounded by the local scale (external factors)?*
– *To what extent do planning innovations involve trade-offs between the social, environmental and economic dimensions of sustainability?*
– *To what extent can planning innovations be transferred from one context to another?*

Selection of city case studies

The case studies in this book have been chosen to directly reflect projects focusing on social sustainability, environment sustainability and urban economic sustainability. Many of the case studies include direct impacts on the social dimension as well as on the environment and also have implications in terms of urban economic sustainability. But in terms of structure, the case studies in the book are grouped into one of the three sustainable planning priorities – social, environmental or economic, according to their principal focus.

Nevertheless the book recognises that the multi-dimensional nature of sustainability means that success along one of sustainability's three dimensions – environmental, social and economic – might not necessarily bring success along the other dimensions. Indeed, success on a given dimension might have to be traded off against outcomes on one or both of the other dimensions. Scott Campbell (1996) argued this caveat in clear terms two decades ago when sustainability issues were expanding rapidly after the Brundtland declaration of sustainability principles, but the issue remains central. In the concluding chapter we return to this issue, drawing on the case studies in the book to suggest the extent to which trade-offs might be required in practice to achieve sustainability 'success.'

Planning innovations are attempts to make changes to status quo practices and situations (Bai et al. 2010, 313). Each chapter features a planning innovation – in the sense of a new kind of intervention – that has the potential to be replicated elsewhere and thus can lead to more sustainable planning practices worldwide or achieve a 'transition' towards more sustainable practices in planning. Innovations can be seen as experiments that require refinements to be adopted and to become mainstream planning practices.

Lastly, it was considered important to include case studies from Latin America and from Asia as cities from those parts of the world are experiencing rapid urbanisation and are thus facing specific sustainability challenges not necessarily found in other parts of the world. As stated by Sorensen et al. (2004), many books that have dealt with sustainability have focused on European and North American experiences. We believe it is essential to also understand the process of implementation of urban sustainability in the Latin American and, in particular, the Asian contexts.

References

Atkinson G., Dietz S., Neumayer E. and Agarwala M. (eds) (2014) *Handbook of Sustainable Development*, Edward Elgar, Cheltenham.

Bai X., Roberts B. and Chen J. (2010) "Urban sustainability experiments in Asia: Patterns and pathways", *Environmental Science and Policy* 13 (4) 312–325.

Baker S. (2016) *Sustainable Development*, Routledge, London.

Bovill C. (2014) *Sustainability in Architecture and Urban Design*, Routledge, London.

Campbell S. (1996) "Green cities, growing cities, just cities? Urban planning and the contradictions of sustainable development", *Journal of the American Planning Association* 62 (3) 296–312.

Campbell S. (2016) "The planner's triangle revisited: Sustainability and the evolution of a planning ideal that can't stand still", *Journal of the American Planning Association* 82 (4) 388–397.

Colantonio A. and Dixon T. (2010) *Urban Regeneration and Social Sustainability: Best Practices from Europe*, Wyley & Blackwell, London.

Darchen S. (2016) "Regeneration and networks in the Arts District. Rethinking governance models in the production of urbanity", *Urban Studies* 54 (15) 3615–3635

Geels F.W. (2011) "The multi-level perspective on sustainability transitions: Responses to seven criticisms", *Environmental Innovations and Societal Transitions* 1 24–40.

Geels F.W. (2014) "Regime resistance against low-carbon transitions: Responses to seven criticisms", *Theory, Culture & Society* 31 (5) 21–40.

Glaeser E. (2011) *Triumph of the City*, Pan Macmillan, London.

Goodland R. (1995) "The concept of environmental sustainability", *Annual Review of Ecological System* 26 1–24.

Harvey D. (1982) *The Limits to Capital* Oxford, Blackwell.

Hirt S.A. (2016) "The city sustainable: Three thoughts on 'Green cities, Growing cities, Just cities'", *Journal of American Planning Association* 82 (4) 383–384.

Hunt M. and De Laurentis C. (2015) "Sustainable regeneration: A guiding vision towards low-carbon transition?", *Local Environment* 20 (9)1081–1102.

Loorbach D. and Rotmans J. (2010) "The practice of transition management: Examples and lessons from four distinct cases", *Futures* 42 237–246.

Markard J., Raven R. and Truffer B. (2012) "Sustainability transitions: An emerging field of research and its prospects", *Research Policy* 41 (6) 955–967.

McCann E. and Ward K. (2008) *Mobile Urbanism: Cities and Policy-making in the Global Age*, University of Minnesota Press, Minneapolis.

Meadowcroft J. (2009) "What about the politics? Sustainable development, transition management and long term energy transitions", *Policy Sciences* 42 (4) 323–340.

Meijer I. and Hekkert M.P. (2007) "Managing uncertainties in the transition towards sustainability: Cases of emerging energy technologies in the Netherlands", *Journal of Environmental Policy & Planning* 9 (3–4) 281–298.

Meijer A. and Thaens M. (2018) "Urban technological innovation: Developing and testing a sociotechnical framework for studying Smart City projects", *Urban Affairs Review* 54 (2) 363–387.

Moore T., Haan F.J. de, Horne R., Gleeson B.J. (eds) (2018) *Urban Sustainability Transitions Australian Cases- International Perspectives*, Springer, Singapore.

Mössner S. (2016) "Sustainable urban development as consensual practice: Post-Politics in Freiburg, Germany", *Regional Studies* 50 (6) 971–982.

Newman P. and Jennings I. (2008) *Cities as Sustainable Ecosystems. Principles and Practices*, Island Press, Washington.

Newman P. and Kenworthy J. (1999) *Sustainability and Cities: Overcoming Automobile Dependence*, Island Press, Washington.

Ooi G.L. (2008) "Cities and sustainability: Southeast Asian and European perspectives", *Asia Europe Journal* 6 (2) 193–204.

Payre B. (2010) "The importance of being connected. City networks and urban government: Lyon and Eurocities (1990–2005)", *International Journal of Urban and Regional Research* 34 (2) 260–280.

Peck J. and Theodore N. (2015) *Fast Policy*, University of Minnesota Press, Minnesota.

Roseland M. (2012) *Towards Sustainable Communities*, New Society Publishers, Canada.

Rutherford J. and Coutard O. (2014) "Urban energy transitions: Places, processes and politics of socio-technical change", *Urban Studies* 51 (7) 1353–1377.

Saha D. and Paterson S.R.G. (2008) "Local government efforts to promote the 'three Es' of sustainable development", *Journal of Planning Education and Research* 28 (1) 21–37.

Savitch H.V. and Kantor P. (2002) *Cities in the International Marketplace: The Political Economy of Urban Development in North America and Europe*, Princeton University Press, Princeton.

Schweitzer L. (2016) "Tracing the justice conversation after 'Green Cities, Growing Cities'", *Journal of American Planning Association* 62 (3) 374–379.

Sorensen A., Marcotullio P.J. and Grant J. (2004) *Towards Sustainable Cities: East Asian, North American, and European Perspectives on Managing Urban Regions*, Ashgate, Aldershot.

Sotarauta M., Beer A. and Gibney J. (2017) "Making sense of leadership in urban and regional development", *Regional Studies* 51(2) 187–193.

Temenos C. and McCann E.A. (2012) "The local politics of policy mobility: Learning, persuasion, and the production of a municipal sustainability fix", *Environment and Planning A* 44 (5) 1389–1406.

United Nations (2015) *Transforming our World: The 2030 Agenda for Sustainable Development*, Department of Economics and Social Affairs, Washington.

Vojnovic I. (2012) *Urban Sustainability: A Global Perspective*. Michigan State University Press, East Lansing.

While A., Jonas A.E.G. and Gibbs D.C. (2004) "The environment and the entrepreneurial city: Searching for the 'sustainability fix' in Leeds and Manchester", *International Journal of Urban and Regional Research* 28 (3) 549–569.

World Bank (1992) *World Development Report 1992: Development and the Environment*, Oxford University Press, New York.

2

CALGARY, CANADA

Policy co-production and indigenous development in urban settings

Yale D. Belanger, Katherine A. Dekruyf and Ryan C. Walker

Introduction

Policy co-production's promise for accommodating Indigenous interests in municipal planning and policy development processes has taken on heightened academic significance in Canada since the release of the Truth and Reconciliation Commission's (TRC) final report in 2015. A number of cities have responded in kind by attempting to incorporate Indigeneity into municipal institutions, reflecting international trends evident in colonial countries such as New Zealand and Australia, where, like Canada, the former embraced reconciliation as a basis for promoting greater Indigenous social and political inclusion. Despite its promise, policy co-production remains conceptual, and no accepted framework is available for local officials looking to engage citizens about issues related to Indigenous governance or Indigenous pre-contact occupancy. As with other cities, the City of Calgary (hereafter the City) has embraced the call to develop meaningful Indigenous-municipal relationships. Building on Porter and Barry's (2015, 37) work, the City was presented with an opportunity in 2012 to expand the boundaries of its planning practice to "accommodate Indigenous interests" based "on terms that are themselves up for negotiation" rather than on "the unquestioned expectation that they must fit into established ways of knowing and acting." The outcome of this policy co-production trial is examined below in greater detail emphasizing specifically those reasons municipal officials used to reject forging working relationships with the urban Indigenous community.

From August 2014-April 2015, 12 personal interviews were conducted with seven municipal officials (mayor, six city councillors) and the heads of five urban Indigenous organizations with strong municipal government ties. Approximately one hour in length, the interviews followed a general format: researchers engaged each participant in a discussion while posing a number of pre-determined questions designed to 1) keep the interviewer attuned to the major themes being investigated while 2) eliciting the participants' stories that, in this instance, act as a source of understanding to provide insight into personal decision-making (Cortazzi 2001). The interviews were digitally recorded and transcribed, and the researchers reviewed and finalized the coding process using NVivo10 software as a supportive tool (Clark and Braun 2014). The interviews were used to "clarify the relations of individuality, both as output and input, to its sociocultural context" while eliciting behaviours and attitudes that suggest "hidden or latent dimensions of the organization of persons and of the sociocultural matrix and their inter-actions" (Levy and Hollan 1998, 334). Data collection and analysis proceeded simultaneously, and transcripts were read and re-read to ensure accuracy and thematic applicability to the original data.

The City: Calgary

Calgary is sited at the confluence of the Elbow and Bow Rivers in Southern Alberta in what is the northern region of the lands the Blackfoot have called home for millennia. Regarded as Blackfoot traditional territory (and Treaty 7 lands), the City is also known as *Moh-kins-tsis*, or 'elbow many houses,' where Indigenous peoples of the pre-European period comprised complex organized societies (Binnema 2004). The settlement era brought the North-West Mounted Police (NWMP) to the region in the 1870s, extending Canadian sovereignty into the west, which led to Fort Calgary's quick construction (Foran and Cavell 1978). The NWMP was tasked with both monitoring Indigenous people recently sequestered to reserves and encouraging their social and economic transition into sedentary farmers and ranchers. This dual-track civilization strategy confounded Indigenous peoples who believed that by signing Treaty 7 in 1877 with the British Crown and Canadian officials they would retain full use of their lands now being shared with settlers (Treaty 7 Elders et al. 1996).

Unfettered access to Indigenous lands stimulated southern Alberta's development, which surged from the 1890s until World War I. During this period the City began its transition into an urban landscape (Melnyk 1985). The City's population grew rapidly during the 20th century, and by 2016, it was Canada's third largest city (1.24 million). Between 2011 and 2016, the Canada Census listed Calgary as the country's fastest growing metropolitan centre. Provincially, the City is Alberta's largest city both in terms of population and municipal territory, resulting in a low-density, mono-centric setting that municipal planning officials envision developing into a polycentric centre. Of the City's total population, 337,420 (28.1 per cent) people belong to a minority group, the top three groups being South Asian, Chinese, and Filipino. A growing immigrant population has led city officials to place more emphasis on promoting Calgary as a diverse and inclusive city. In 2016, the City's Indigenous population was 35,195, representing 13.6 per cent of the provincial Indigenous population and 2.8 per cent of the City's population. Calgary's Census Metropolitan Area (CMA) includes the Tsuu T'ina Nation, a First Nations community of roughly 2,000 individuals.

The urban sustainability challenge: Integrating indigeneity

Municipal resistance to integrating urban Indigenous peoples into political decision-making and planning regimes is common in Canada, and it is customary in Australia and New Zealand, to name two colonial countries confronting similar trends. Municipal resistance is fuelled by what has been described as municipal colonialism, a city-planning approach that dates to at least the 1920s (Stanger-Ross 2008; Porter 2013). While municipal colonialism may be conceptually specific to Canada, comparable processes of denying Indigenous social, economic, or political claims to the city are unmistakable in Australia and New Zealand. Municipal leaders in Canada referenced their nation-state's Indigenous policy-making models, and in a similar fashion would structure "political realities and subversive political imaginaries," where authority and legitimacy were assigned in ways that benefitted "some citizens systematically and, just as systematically" disempowered and harmed other citizens (Biolsi 2005, 240–241). A network of cities and towns similarly emerged in colonial countries whose leaders framed Indigeneity and urbanism as first and foremost incommensurable (Peters 1996; Maaka and Fleras 2005).

In April 2014, the City's Mayor Naheed Nenshi publicly announced the next 12 months would be considered a Year of Reconciliation to ensure "our Aboriginal population has a meaningful role within our community, as full and equal participants in our city's quality of life." Prior to this juncture disruptive periods of animosity characterized municipal-Indigenous interaction (e.g., Wood 2003). Consequently urban Indigenous leaders considered the reconciliation proclamation to be more than a symbolic gesture, and most believed that they were on the cusp of fashioning meaningful interpersonal and political relationships with non-Indigenous peoples that could lead to greater local participation in municipal policy production (Porter 2013; Walker and Belanger 2013; Alcantara and Nelles 2016). Buttressing these beliefs were numerous Supreme Court of Canada (SCC) decisions providing clarity concerning urban Indigenous peoples' rights (Belanger 2013), about municipal governments' role as an administrative body of the crown, and the legal duty to consult's potential urban function (MacCallum and Viswanathan 2013). Indigenous peoples then referenced the United Nations Declaration on the Rights of Indigenous Peoples (UNDRIP) to help contextualize these rights, specifically self-determination rights (Belanger 2011).

Despite the publicly stated desire to work with Indigenous people the mayor and the city council were unable to generate policy co-production with the urban Indigenous community. Arguably, policy co-production remains difficult for municipal officials to comprehend for it represents a radical departure from liberal democratic ideals promoting individual equality. For many municipal officials this is an imposing step due primarily to their reluctance to accept that Indigenous peoples are entitled to a meaningful measure of self-determination based on prior occupancy rights (Kymlicka 2001). Policy co-production has been attempted in several Canadian cities, but the results to date have been less than inspiring (Skelton 2002; Belanger and Walker 2009; Walker and Belanger 2013). As Walker et al.'s

(2011, 193) study of co-production of various policy fields in four Manitoba cities concluded (see also Alcantara and Nelles 2016), which speaks to the majority of national efforts to date:

> governments are not really co-producing policy with Aboriginal communities at all: they are simply striking broad-based advisory 'tables', with lots of 'voices', to assist with the implementation of government policy that has been derived from agenda-setting onwards (until the implementation stage).

What is intriguing is that municipal leaders appear *willing to consider* adopting an augmented role for urban 'Indians' (Hanselmann 2001; Papillon and Juneau 2015). These efforts tend to break down at the conceptual level (and as will be argued below the ideological level). That is, municipal officials and urban Indigenous leaders for the most part worked well together as they seek improved interactions. Nevertheless, municipal officials tend to balk at trying to build relationships due to a lack of understanding of: 1) urban Indigenous self-determination as an institution; and 2) Indigenous peoples' municipal status as residents. The latter issue is somewhat confusing when one considers that urban Indigenous peoples are municipal citizens in the same way that non-Indigenous peoples are citizens. It nevertheless remains a source of concern, as does the debate concerning urban Indigenous self-determination. Capturing the essence of this municipal uncertainty is a popular and influential question circulating through municipal government channels concerning Indigenous self-determination: how is it that Indigenous peoples, who were expected to disappear physically and/or as a policy community, evolved into self-determining agents? The fact that this question persists helps us to better understand why Indigenous claims to urban self-determination or self-governance are not applauded. Rather they are considered an overt challenge to municipal histories that continue to reject urban-Indigenous compatibility. Consequent Indigenous claims reinforce municipal attitudes equating urban Indigenous peoples as a problem, one of municipal colonialism's original founding tenets (Newhouse and FitzMaurice 2012).

A key message to emerge from this debate is that urban Indigenous peoples cannot be considered integrated as political partners, even though municipal officials appear willing to engage Indigenous peoples as political partners. Policy co-production in this setting takes on heightened significance due to its potential to help resolve this observable tension: it advances municipal-urban Indigenous relationships through "policy generation and implementation process where actors outside of the government apparatus are involved in the creation of policy, instead of only its implementation" (Belanger and Walker 2009, 120). Policy co-production further encourages government and community-based actor interface from problem/issue identification and priority setting to program and service delivery, and beyond. Co-production's benefits can include sharing responsibility for issue definition and priorities identification to assigning public resources to programs that are likely to achieve positive outcomes. Local government(s) arguably must forsake the current practice of refusing to work with suitable urban Indigenous leaders, experts, and community members (many of these individuals will present longevity beyond any single civic executive, administration or budget year).

Drawing from these insights and Newman's (2014, 195) work, the following section evaluates the Paskapoo Slopes development consultation process discussed below that pursued policy co-production according to four criteria: 1) process, or whether "public policy is converted into substantial legislation or public programming"; 2) goal attainment, of the combined goals where these policy objectives were achieved; 3) distributional outcomes, which evaluates how certain groups benefit from specific policies in particular ways; and 4) political consequences, which identifies how governments and policy actors benefit from the "public reaction to, or perception of, a policy." After aggregating these various elements, we produce conclusions that highlight the planning innovation's (policy co-production) existing procedural strengths and weaknesses, with the goal of pinpointing where positive change can occur resulting in improved future consultation.

The planning innovation: Policy co-production

Confronted with a changing social, economic, and legislative environment, Mayor Nenshi, the City council, and urban Indigenous leaders in the mid-2010s acknowledged the inevitability of political interaction. This in part led to the reconciliation proclamation and the dedicated one-year period for enhancing urban Indigenous political contributions. Not all city councillors were convinced of Mayor Nenshi's approach, but his combined popularity and doggedness convinced several influential councillors of its validity. Mayor Nenshi's sincerity further convinced urban Indigenous leaders to intensify their lobbying efforts to participate in municipal policy-making. By 2015 it appeared as though the stage was set to test the City's confidence in working with urban Indigenous peoples when an experiment materialized in the form of a development project directed by Trinity Development Group Inc. (established in 1992), which was at the ground-breaking stage.

Innovation process

Trinity is a commercial real estate development company specializing in urban mixed-use centres, community centres, and large format centres. In 2012, it announced plans to build a 100 acre urban ski village called Calgary's Whistler that would offer a mix "of retail shops, office space and homes [on] the foot of the Paskapoo Slopes" (Markusoff 2014). Concerns were raised about the potential environmental impact and the anticipated traffic problems that could lead to significant congestion at the intersection of two major arteries. Trinity altered its plans to be more "pedestrian-friendly, more mixed use" and agreed to donate 160 acres to the City of Calgary for a protected regional park. Indigenous elders observing the development's evolution opposed these changes for fear that construction would endanger a historically vital "buffalo kill site" they wanted protected.

City planners approached the elders and members of the Calgary Aboriginal Urban Affairs Committee (CAUAC) and worked to forge a new proposal sporting 16 amendments that the Calgary Planning Commission approved in July 2015. The new proposal restricted development to one-third of the area with the remaining two-thirds designated for a new regional park (CBC News 2015).

Unfortunately, while the amendments catalogued the elders' concerns they lacked institutional force needed to compel formal changes. But significant Indigenous input was largely absent as the development progressed. That is, until it came time to name the community. City officials and Blackfoot elders met to discuss potential names, but a majority of city councillors opposed the proposed name (*Aiss ka pooma*) on the grounds that it was too intricate a pronunciation for non-Indigenous Calgarians. The name Medicine Hill was later proposed but to little fanfare (Fletcher 2016). One councillor in particular identified the need to balance practicality with Indigenous respect, and voted for the commission's recommendation that four streets be named with Blackfoot words. Another openly rejected this proposal: "[t]his is going beyond politically-correct [and] it's totally against the policy that names be easily pronounced" (Kauffman 2016). In the end, the street naming exercise proceeded while Indigenous input into the development's other features remained severely constrained.

Reflecting on Newman's (2014) four measurement criteria, the lack of goal attainment is notable. Municipal and Indigenous leaders may have agreed to consultations, but their representatives failed to identify key issues or how to work with one another. Policy co-production encourages problem identification among municipal and Indigenous leaders, after which one can establish priorities and craft outcomes based on sharing responsibility and resource distribution (Walker and Belanger 2013). In this example, partner discussions were reduced to a street naming exercise even though city officials appeared to be motivated by Mayor Nenshi's interest in ensuring social justice for Indigenous peoples and greater levels of social inclusion (Dekruyf 2017). This at the same time is a minor victory considering the animosities that developed between the City and the Tsuu T'ina Nation in the early 2000s after negotiations broke down regarding a ring road extension to be built on reserve land needed to accommodate the city's southwestern residents (Wood 2003). Intermittent attempts to return to the table occurred, but city officials largely steered clear of working with Indigenous peoples, which led to First Nations and urban Indigenous feelings of political marginalization. Mayor Nenshi's actions, specifically his reconciliation proclamation, intimated that for the first time in a decade urban Indigenous leaders were considered to be political equals to the non-Indigenous community.

Indigenous respondents were not deterred from pursuing policy co-production despite its failure to produce substantive outcomes. They did however suggest that reconciliation should ground policy co-production efforts thus offering: 1) a "decolonizing place of encounter between settlers and Indigenous people"; 2) "space for collective critical dialogue"; and 3) "a public remembering embedded in ethical testimonial, ceremonial, and commemorative practices" (Regan 2010, 12).

Indigenous leaders concluded that reconciliation needed to accommodate Indigenous peoples' "aspirations in relation to land and political autonomy" (Short 2008, 5). The City's vision of reconciliation was conspicuously less complex: municipal officials' chose instead to emphasize managing stakeholder engagement (Laforest and Dubois 2017). Indigenous leaders claimed that the "hegemony of the settler myth," described by Miller (2017, 267) as the process of assessing non-Indigenous political and economic progress according to a belief in Indigenous inferiority, grounded the municipal approach to co-production. Partnerships in this milieu consequently could not develop for urban Indigenous people were not considered rights holders, or for that matter political equals. They were instead stakeholders to be managed, which meant that city officials never took the time to effectively reflect on the source of Indigenous grievances. To do otherwise would have obliged municipal officials to formally acknowledge colonialism's historic certainty and its ongoing contemporary municipal political influence.

This lack of common policy outcomes compromised the consultations, as did municipal officials retaining the final say on the development's progress, and whether the land would be afforded municipal protections. According to Newman (2014), then, public policy was not developed: nothing emerged that could be converted into substantial legislation or public programming. What we are left to examine, therefore, are the political consequences linked with how public reaction to, or perception of, a policy influences governments and policy actors alike. In this regard, the non-Indigenous community paid little attention to the entire event. Support was neither forthcoming nor was it formally withheld, suggesting civic indifference. This demands further research, specifically examining why such attitudes continue to present. As a result, our next best course of action is to explore the Indigenous response to the consultations. The analysis that follows is informed by two key reasons municipal officials identified for branding urban Indigenous peoples as stakeholders, rather than as political equals with self-determination rights, both of which are undergirded by the "hegemony of the settler myth." The first: local officials claimed that their desire to ensure citizen equality restricted how they could engage Indigenous leaders. Second, our data verified that local officials identified the urban Indigenous community as lacking political legitimacy.

In the first instance, municipal officials acknowledged (with no sense of irony) the ongoing influence of colonial attitudes in urban Indigenous socio-economic and political marginalization. Although each one agreed that there was a need to build inclusive cities and neighborhoods, most were unable or unwilling to formulate alternative programs to help elevate Indigenous peoples. Their apprehension was driven by local concerns that municipal-Indigenous engagement could be interpreted as the city council privileging an ethno-cultural group over mainstream Canadians. For example, during the interviews the idea of race-based rights and special privileges generated markedly negative responses from city councillors. This led to a hypothesis that drove municipal response strategies: *if* Indigenous peoples *are* indeed municipal citizens they are not entitled to special rights. Yet municipal officials simultaneously portrayed urban Indigenous peoples as municipal outsiders (temporary residents but not citizens), which brings us to the second point: urban Indigenous self-determination claims are considered legally tenuous. That is, one cannot be self-determining or self-governing devoid of a sovereign land base. Indigenous claims to self-determination or self-government were therefore dismissed according to the understanding that municipal citizens cannot be considered self-determining outside of *Municipal Government Act* strictures or Canadian law. Finally, local officials were adamant that authentic Indigenous identities and self-determination rights are nested within First Nation communities and as such are not transferable to the city. Such conclusions challenge several SCC decisions classifying urban Indigenous peoples as unique political communities whose residents may also retain First Nations membership (Belanger 2013).

In sum, municipal officials refused to acknowledge urban Indigenous political legitimacy or the UNDRIP's certainty in urban Indigenous self-determination. Consequently, urban Indigenous populations are erratically portrayed as members of self-governing First Nations (i.e., band members living in the city); as members of non-citizen, special interest groups federal officials retain exclusive responsibility for; or, as citizens with the same rights and privileges afforded all other Canadians (Belanger and Dekruyf 2017). As Daigle (2016, 267) describes it, "recognition-based strategies are founded on and materially reproduce colonial imaginaries of territory that continue to inflict violence on Indigenous legal and governance orders while facilitating the economic and political sovereignty of Canada." An institutional invisibility results, local officials responsible for planning for urban Indigenous peoples remain uninformed about urban Indigenous concerns or needs, and this undermines urban Indigenous claims to a right to the city (e.g., Winders 2012; Belanger and Dekruyf 2017).

The vast majority of Indigenous individuals interviewed expressed frustration that the City remains a colonial space characterized by inferior Indigenous socio-economic outcomes (employment, health, education). They further emphasized the paradox of Calgary officials' recognition that the city rests on traditional Blackfoot territory and their coincident denial of substantive urban Indigenous self-determination or citizenship rights. In this regard, it was not surprising that the project participants identified municipal institutions as poorly matched to their needs. Whereas project participants appreciated the City's willingness to accept their collective voice as an ad hoc advisory body to council, and were generally satisfied with the perceived degree of inclusion, one participant noted that the City of Calgary did little more than create a "little site . . . for Aboriginal people." One councillor confirmed that urban Indigenous peoples are commonly an afterthought, admitting that "it often feels a little bit more token than anything else . . . I know we're trying to get better at it, but I don't think that we do it very well."

Despite such promising beginnings, few substantive outcomes came out of the Paskapoo Slopes negotiations – at least from an urban Indigenous perspective. The City had no formal structures for engaging the urban Indigenous community, which is simultaneously a social and an institutional issue: 1) it is reflective of municipal colonial attitudes about urban Indigenous peoples that have become embedded in municipal policies; and 2) municipal opposition to working outside the current system's boundaries to devise new engagement models results. The need to counter municipal colonialism by promoting the mutual creation of inter-active strategies thus forging a base for policy co-production fuelled the urban Indigenous desire for a seat at the planning table. City officials were however unwilling to concede to any Indigenous demands, relying instead on the familiarity of post-colonial institutions (attitudes, policy, and governance). This institutional foundation combined with the need to ensure regional economic development meant that accepting the legitimacy of urban Indigenous land rights was not viable. Even when it appeared that progress was being made, the City still maintained control over how development would proceed, and accepted only symbolic input from the Indigenous community in the form of naming four streets. Municipal resistance undergirded by a colonial ethic demanding Indigenous peoples 'prove their worth' did not permit meaningful Indigenous political participation, which in turn undermined policy co-production's innovative nature. This is a harsh realiza-tion in what is Canada's self-proclaimed reconciliation era.

Conceptually co-production remains a positive approach that has not been effec-tively harnessed by the City, and the implications associated with these ongoing setbacks demands a brief word. As noted, municipal officials embrace a liberal view of social cohesion that is grounded by a belief in individualism augmented by hard work, and a sense of universalism. Speaking to individualism, Indigenous peoples forming communal societies are considered suspect and remain outside the politi-cal imaginary until they can prove themselves ready for integration. Indigenous demands for what Calgary's leaders consider special recognition of race-based rights are considered anathema to political universalism, even if both the federal government and the Supreme Court of Canada have formally recognized these rights as existing. The resulting impasse is therefore ideologically grounded – and the implications for remaining ideologically entrenched are significant.

The City may choose to maintain the status quo and proceed unilaterally with local development or by periodically permitting symbolic urban Indigenous input, as with the Paskapoo Slopes. Whereas the latter approach may in the short term satisfy urban Indigenous leaders, eventually they will demand a more formidable role in local affairs. But pursuing a strategy of unilateral municipal policymaking will inflame tensions. In either case City officials continue to portray urban Indigenous peoples as a special interest group. It is important to note that urban Indigenous leaders in the Paskapoo Slopes negotiations did not employ all of the available tools to contest municipal actions, resistance being the most prominent method. Perhaps the Paskapoo Slopes project was considered too inconsequential to dispute, or maybe forging working relationships with the City was considered too important an outcome to risk to contentious action. As Tully (1995, 23) has discussed, utiliz-ing resistance for purposes of enhancing agency is an established historic Indigenous strategy, and Indigenous peoples have a range of repertoires adaptable to the munic-ipal setting. Litigation is one such strategy, and will in all likelihood be pursued should Calgary choose to move forward arbitrarily in the future. It is an expensive and time-consuming alternative, but pursuing litigation can compel the courts and federal officials to establish contours of meaning for the idea of the inherent right to urban Indigenous self-government. Such decisions may also dictate that the City needs to formally respond to demands for Indigenous self-determination currently ignored in contemporary interactions.

Just as notable, and likely a more threatening posture, is the potential for Indigenous direct action and protest activity. Caledonia nearby Hamilton Ontario is the most recent high-profile example of a municipality being forced to contend with Indigenous protest, but as demonstrated by this and other events, direct action and the risk of blockades and occupations are proven tactics. They can drive away tourist traffic, lead to business closures, and set back if not freeze local development for months at a time, all the while irreparably damaging local relationships in certain cases (e.g. Belanger and Lackenbauer 2014). An occupation at the Paskapoo Slopes site similar to that which occurred at Caledonia in 2006 (and continues in distinct iterations) could have effectively halted development. The urban Indigenous community would likely claim a victory by threatening the economic prosperity of several Calgary wards, but they would likely destabilize social relations. Admittedly Calgary's urban Indigenous community is informally organized, and for now this makes fashioning coherent resistance efforts directed at municipal projects a rather difficult proposition. This will not remain the case and needs to be acknowledged.

The trade-offs related with maintaining a strictly liberal approach to the detriment of developing a more inclusive meaning of social cohesion embracing Indigenous peoples are at once social and economic. Although we lack the empirical data to determine precisely the costs of these and similar actions, the current trajectory of City officials governing over urban Indigenous peoples who have recognized rights of self-government, which they hypothesize should lead to formal dialogues developing between political equals as opposed to symbolic moments of consultation between antagonists, will prove unsustainable, economically costly, and socially unmanageable. The potential costs of inaction associated with maintaining the status quo have been identified. Whether the City is able and/or willing to reconcile evolving Indigenous rights with a public conviction in liberal beliefs emphasizing the individual and universalism remains the important question to resolve if policy co-production is to have any positive future impact in Calgary.

Conclusion

Municipal-Indigenous relationships are essential resources that must be nurtured to ensure sustainable municipal development. Hence the frustration concerning the reality of policy co-production's failure in Calgary: it is ideologically anchored by historic social attitudes that portray Indigenous people as rural and communal people out of place living in the city. This complicates any and all proposed co-production models ostensibly predicated on equals coming together to mutually impart concessions upon their negotiating partners resulting in both sides giving a little to gain a little. The municipality was unwilling to concede, and chose to rely on its own internally crafted policies that were underscored by an ideological certainty in Indigenous inferiority to justify its opposition. The municipal officials we interviewed were unable to break away from the idea that Indigenous peoples are stakeholders as opposed to full, equal, and self-determining partners as policy co-production intended, and a theme that demands additional study. Nor do they consider urban Indigenous peoples as possessing full citizenship rights or UNDRIP-recognized rights of self-determination. All the same municipal leaders acknowledge Calgary is sited on Blackfoot territory, and that the ensuing local development was and in many ways remains harmful to Indigenous peoples.

The Paskapoo Sites example is illustrative of similar attempts in various Canadian cities to co-produce policy. As Alcantara and Nelles (2016) have shown, although Indigenous–municipal political relationships are in many ways more progressive than those that continue to develop at higher governmental levels, municipal governments lack the necessary knowledge, expertise, resources, and at times willingness, to successfully engage Indigenous issues, partners, and constituents. The general lack of progress is similar to other colonial countries such as New Zealand (e.g., Keiha and Moon 2008) and Australia (e.g. Porter 2013), for instance, where existing constraints to collaboration undermine potentially transformative initiatives at the local government level. Municipal resistance to Indigenous political participation remains grounded by colonial beliefs that Indigenous peoples are separate and politically unproven. The Paskapoo Slopes negotiations represent a brief moment of co-production that unfortunately failed to lead to joint management of the contested space. This was a lost opportunity and a moment that the City exploited to defuse an Indigenous movement seeking to reclaim urban territorial sovereignty.

Policy co-production's potential is evident, but what steps are needed to move beyond the conceptual to the operational? Municipal leaders have at the very least accepted that community development is contingent on Indigenous support. In turn this demands establishing local institutions promoting relationship building to mitigate the impacts of the types of systemic problems identified in this chapter. To date symbolic gestures represent the substance of change – the City flies a Treaty 7 flag at city hall and officials acknowledge that Calgary sits on Indigenous lands. Institutional changes are all the same required to ensure that Indigenous participation develops from more than token gestures employed to defuse tensions emerging from urban Indigenous land claims. Recently, the City and CAUAC produced the *White Goose Flying Report* (CAUAC 2016), which is a reconciliation scaffold. This is a key launching point toward forging institutional frameworks clarifying the various parties' roles and responsibilities, and how to pursue working relationships. Such a course of action depends obviously on municipal officials jettisoning the colonial baggage that ultimately stymied the potential for change linked with the Paskapoo Slopes development. Until this happens and Indigenous peoples are elevated from the position of stakeholders to right-holders, City of Calgary officials will continue to prescribe to them an advisory role, the implications of which will remain considerable for municipal development.

References

Alcantara, C., and Nelles, J. (2016). *A Quiet Evolution: The Emergence of Indigenous-Local Intergovernmental Partnerships in Canada*, University of Toronto Press, Toronto.

Belanger, Y. (2011) "The United Nations Declaration on the Rights of Indigenous Peoples and Urban Aboriginal Self-Determination in Canada: A Preliminary Assessment", *Aboriginal Policy Studies* 1, 1, 132–161.

Belanger, Y. (2013) "Breaching Reserve Boundaries: Canada v. Misquadis and the Legal Evolution of the Urban Aboriginal Community". In E. J. Peters and C. Andersen (eds), *Indigenizing Modernity: Indigenous Identities and Urbanization in International Perspective*, UBC Press, Vancouver, 68–69.

Belanger, Y., and Dekruyf, K. (2017) "Neither Citizen nor Nation: Urban Aboriginal (In) visibility and Co-production in a Mid-sized Southern Alberta City", *Canadian Journal of Native Studies* 37, 1, 1–28.

Belanger, Y., and Lackenbauer, P. (eds) (2014) *Blockades or Breakthroughs? Aboriginal peoples Confront the Canadian State*, McGill-Queen's University Press, Kingston and Montreal.

Belanger, Y., and Walker, R. (2009) "Interest Convergence & Co-production of Plans: An Examination of Winnipeg's 'Aboriginal Pathways'", *Canadian Journal of Urban Research* 18, 1, 118–139.

Binnema, T. (2004) *Common and Contested Ground: A Human and Environmental History of the Northwestern Plains*, University of Toronto Press, Toronto.

Biolsi, T. (2005) "Imagined Geographies: Sovereignty, Indigenous Space, and American Indian Struggle", *American Ethnologist* 32, 2, 239–259.

CAUAC (Calgary Aboriginal Urban Affairs Committee) (2016) *White Goose Flying: A Report to Calgary City Council on the Indian Residential School Truth and Reconciliation Calls to Action 2016*, Calgary Aboriginal Urban Affairs Committee, Calgary.

CBC News (2015) "Paskapoo Slopes development passes first hurdle at city hall" (www. cbc.ca/news/canada/calgary/paskapoo-slopes-development-passes-first-hurdle-at-city-hall-1.3164446) Accessed 13/05/2018.

City of Calgary (2014) "Proclaiming the Year of Reconciliation" (http://calgarymayor.ca/ stories/proclaiming-the-year-of-reconciliation) Accessed 13/05/2018.

Clarke, V., and Braun, V. (2014) "Thematic analysis". In A. C. Michalos (ed.), *Encyclopedia of Quality of Life and Well-Being Research*, Springer, New York, 57–71.

Cortazzi, M. (2001) "Narrative Analysis in Ethnography". In P. Atkinson (ed.), *Handbook of Ethnography*, Sage, London, 384–394.

Treaty 7 Elders and Tribal Council, Hildebrant, W., Carter, S., and Rider, D. F. (1996) *The True Spirit and Intent of Treaty 7*, McGill-Queen's, Kingston and Montreal.

Daigle, M. (2016) "Awawanenitakik: The Spatial Politics of Recognition and Relational Geographies of Indigenous Self-determination", *The Canadian Geographer / Le Géographe canadien* 60, 2, 259–269.

Dekruyf, K. (2017) "Citizens Minus? Urban Aboriginal Self-Determination and Co-Production in the City of Calgary". MA dissertation, University of Lethbridge, Lethbridge.

Fletcher, R. (2016) "Medicine Hill, not Aisska pooma, selected as new community name" (.www. cbc.ca/news/canada/calgary/paskapoo-slopes-medicine-hill-aiss-ka-pooma-1.3479084) Accessed 13/05/2018.

Foran, M., and Cavell, E. (1978) *Calgary: An Illustrated History*, James Lorimer and Co., Toronto.

Hanselmann, C. (2001) *Urban Aboriginal People in Western Canada: Realities and Policies*, Canada West Foundation, Calgary.

Kaufmann, B. (2016) "Beyond politically correct: Tug of war erupts over difficult-to-pronounce Blackfoot street names", *Calgary Herald* (http://calgaryherald.com/news/local-news/beyond-politically-correct-tug-of-war-erupts-over-difficult-to-pronounce-blackfoot-street-names) Accessed 13/05/2018.

Keiha, P., and Moon, P. (2008). "The Emergence and Evolution of Urban Maori Authorities: A Response to Maori Urbanisation", *Te Kaharoa* 1, 1–17

Kymlicka, W. (2001) "The New Debate Over Minority Rights". In R. Beiner and W. Norman (eds), *Canadian Political Philosophy: Contemporary Reflections*, Oxford University Press, Toronto, 159–176.

Laforest, G., and Dubois, J. (2017) "Justin Trudeau and 'reconciliatory federalism'", *Policy Options* (http://policyoptions.irpp.org/magazines/june-2017/justin-trudeau-and-reconciliatory-federalism/).

Levy, R., and Hollan, D. (1998) "Person-Centered Interviewing and Observation". In H. R. Bernard (Ed.), *Handbook of Methods in Cultural Anthropology*, Walnut Cree, Altamira, 333–364.

Maaka, R., and Fleras, A. (2005). *The Politics of Indigeneity: Challenging the State in Canada and Aotearoa New Zealand*, University of Otago Press, Dunedin.

MacCallum, F. and Viswanathan, L. (2013) "The Crown Duty to Consult and Ontario Municipal-First Nations Relations: Lessons Learned from the Red Hill Valley Parkway Project", *Canadian Journal of Urban Research* 22, 1–19.

Markusoff, J. (2014) "Developer pitches 'Whistler Village' concept for Paskapoo Slopes", *Calgary Herald* (http://calgaryherald.com/news/local-news/developer-pitches-whistler-village-concept-for-paskapoo-slopes) Accessed 25/06/17. Melnyk, B. (1985) *Calgary Builds: The Emergence of an Urban Landscape, 1905–1914*, Canadian Plains Research Center, Regina.

Miller, J. R. (2017) *Residential Schools and Reconciliation: Canada Confronts Its History*, University of Toronto Press, Toronto.

Newhouse, D., and FitzMaurice, K. (2012) "Introduction". In D. Newhouse, K. FitzMaurice, T. McGuire-Adams, and D. Jette (eds) *Well-being in the Urban Aboriginal Community: Fostering Biimaadiziwin, a National Research Conference on Urban Aboriginal Peoples*, Thompson Educational Publishing, Toronto, ix–xxi.

Newman, J. (2014) "Measuring Policy Success", *Australian Journal of Public Administration* 73, 2, 192–205.

Papillon, M., and Juneau, A. (eds) (2015) *Canada: The State of the Federation 2013: Aboriginal Multi-Level Governance*, McGill-Queen's University Press, Kingston and Montreal.

Peters, E. (1996) "'Urban' and 'Aboriginal': An Impossible Contradiction?". In J. Caulfield and L. Peake (eds), *City Lives and City Forms: Critical Research and Canadian Urbanism.* University of Toronto Press, Toronto, 47–62.

Porter, L. (2013) "Coexistence in Cities: The Challenge of Indigenous Urban Planning in the Twenty-First Century". In R. Walker, T. Jojola, and D. C. Natcher (eds), *Reclaiming Indigenous Planning*, McGill-Queen's University Press, Kingston and Montreal, 283–310.

Porter, L., and Barry, J. (2015) "Bounded Recognition: Urban Planning and the Textual Mediation of Indigenous Rights in Canada and Australia", *Critical Policy Studies*, 9, 1, 22–40.

Regan, P. (2010) *Unsettling the Settler Within: Indian Residential Schools, Truth Telling, and Reconciliation*, UBC Press, Vancouver.

Short, D. (2008) *Reconciliation and Colonial Power: Indigenous Rights in Australia*, Ashgate Publishing Limited, Farnham.

Skelton, I. (2002) "Residential Mobility of Aboriginal Single Mothers in Winnipeg: An Exploratory Study of Chronic Moving", *Journal of Housing and the Built Environment* 17, 2, 127–144.

Stanger-Ross, J. (2008) "Municipal Colonialism in Vancouver: City Planning and the Conflict Over Indian Reserves, 1928–1950s", *Canadian Historical Review* 89, 4, 541–580.

Tully, J. (1995). *Strange Multiplicity: Constitutionalism in an Age of Diversity*, Cambridge University Press, New York.

Walker, R., and Belanger, Y. (2013) "Aboriginality and Planning in Canada's Large Prairie Cities". In R. Walker, T. Jojola, and D. C. Natcher (eds), *Reclaiming Indigenous Planning*. McGill-Queen's University Press, Kingston and Montreal, 193–216.

Walker, R., Moore, J., and Linklater, M. (2011) "More than Stakeholders, Voices and Tables: Towards Co-Production of Urban Aboriginal Policy in Manitoba". In E. J. Peters (ed.), *Urban Aboriginal Policy Making in Canadian Municipalities*, McGill-Queen's University Press, Kingston and Montreal, 160–201.

Winders, J. (2012) "Seeing Immigrants: Institutional Visibility and Immigrant Incorporation in New Immigrant Destinations", *The Annals of the American Academy of Political and Social Science* 641, 1, 58–78.

Wood, P. (2003) "A Road Runs Through It: Aboriginal Citizenship at the Edge of Urban Development", *Citizenship Studies* 74, 463–479.

3

HELSINKI, FINLAND

Social sustainability of urban settings – contextually sensitive, participatory approach utilizing PPGIS methodology

Marketta Kyttä, Maarit Kahila-Tani and Anna Broberg

Introduction

Current urban planning strategies favour high densities that can support sustainable modes of transport, efficient energy distribution and the possibility to direct new buildings to infill and brownfield sites instead of greenfield sites (Jabareen, 2006; Jenks, 2010; Newman, 2006; VandeWeghe and Kennedy, 2007). Although many of these ecological benefits are undeniable, the social outcomes of dense urban environments seem to be highly complex and even contradictory (Burton, 2000; Dempsey et al., 2009; Yang, 2008).

With the Helsinki City Plan process that was carried out between 2013 and 2016, the main focus was the infill development of the existing urban structure. The City Plan aimed to rise to the challenges regarding population growth, affordable housing shortage, transportation connections and economic development, among other themes. Urban densification was seen as a key strategy to develop an ecologically, economically and socially sustainable future for the city of Helsinki.

In Helsinki, planners and policy-makers shared a concern about the residents' attitudes toward densification. Traditionally, urban infill projects have been among the topics that accelerate suspicion and mistrust among participants, and cause conflicts (Zhang and Fung, 2013). We had an opportunity to work with the city of Helsinki first in a research project "Urban Happiness" that studied the prerequisites of socially sustainable urban environment (Kyttä et al., 2013, 2016). The research produced experiential, place-based knowledge from residents about the perceived quality of the environment that can be used as base information in context-sensitive urban infilling. The project also provided diagnostic knowledge about the complex relationships between urban structural characteristics and experiential, behavioural and wellbeing outcomes. We were able to identify contextual variation in the ways urban density contributed to social sustainability. The findings have potential practical value by providing planners with context-sensitive information, which is not normally available, that can be used to develop unique, local strategies for urban infill projects.

When the city of Helsinki began the planning process for a new City Plan in 2013, we were able to collaborate with the planners and contribute to the ambitious public participation process of the city (Kahila et al., 2016). To involve residents and stakeholders in the early phase of the planning process, a variety of public participation methods were used, such as seminars, workshops, displays at the City Planning Fair, surveys and meetings. Among them was an online Public Participation GIS (PPGIS) survey that was used during the beginning of the process and that reached nearly 4000 participants. In this survey, residents were, for example, profiled according to their attitudes towards Helsinki's densification policy and were asked to map locations suitable for new building and green areas that should be protected.

Together these two projects paved the way toward diminishing social resistance to urban densification. They produced scientific knowledge, which enhanced an understanding that the same recipes towards a socially sustainable environment and the health and happiness of residents do not apply everywhere.

The city: Helsinki

Helsinki, the capital of Finland, is located in Northern Europe and is one of the fastest growing urban regions in Europe (Turok and Mykhnenko, 2007). In 2017, the Helsinki metropolitan area had about 1.5 million inhabitants and the population is expected to grow to nearly 2 million by 2050 (Vuori and Laakso, 2017). The documented growth extends both outward on unbuilt land (see EEA, 2006) and inward within existing built-up areas (see Jaakola and Lönnqvist, 2007).

A considerable share of the building stock of the suburbs, including also the housing areas dominated by the middle class, consists of apartment blocks built during the post-war era. In terms of density, the city centre neighbourhoods and suburban settings vary between 33 and 134 housing units per hectare. These density levels are relatively high when compared to the Congress for New Urbanism's recommendation (LEED-ND, 2009) of 17 housing units per hectare as the minimum level of density for sustainable neighbourhood development. Still, Helsinki is among the most sparsely built cities in Europe (Kasanko et al., 2006).

The Finnish Land Use and Building Act (2000) increased the autonomy of cities in the field of city planning while it deceased the role of national policy instruments. The law provided also some flexibility for cities to arrange the governance of land use planning. In Finland, cities have zoning monopoly and the city of Helsinki is an especially powerful actor because it owns the majority of land. The Land Use and Building Act obliges cities to realize participatory planning. The city of Helsinki has invested in developing these practices, for example by having a participatory planning unit. Currently, the city is developing even more innovative governance practices and non-governmental actors are increasingly involved in the planning, design and management of Helsinki.

The urban sustainability challenge: Can urban densification promote social sustainability?

According to the existing research literature, the social acceptance of urban density is far from straightforward. For example, McCrea and Walters (2012) noticed that, depending on the context, inhabitants associated densification with both positive and negative effects. Arguably, when well-planned and carefully realized, urban density can meet both the ecological goals for efficient urban structure and also support social sustainability. In the Urban Happiness project we studied this theme empirically in Helsinki city centre and suburban settings, where urban density varied (Kyttä et al., 2016).

Research literature offers an abundance of definitions on social sustainability (Chiu, 2003). For example, Bramley et al. (2009, 2010; also Dempsey et al., 2009) proposed that social sustainability consists of two main dimensions: accessibility (social equity) and experiential outcomes (sustainability of community).[1] Accessibility refers to the equality of access to services and opportunities: essential local services such as shops, schools and health centres; recreational opportunities and open space; public transport; job opportunities; and finally, affordable housing. Experiential outcomes include several themes: pride and attachment to the neighbourhood, social interaction, safety or security, perceived quality of local environment, satisfaction with the home, stability and participation in civic activities.

We found these two dimensions very relevant and agree that treating wellbeing as a mediated outcome is more fruitful than as a component of social sustainability. In the Urban Happiness study, we hypothesized that the two dimensions of social sustainability may also convey a mediating effect on the association between the built environment and wellbeing. We operationalized accessibility as mean distance to everyday services and to personally meaningful places. Experiential outcomes were measured as perceived environmental quality that included four dimensions: social and functional quality, atmosphere and appearance (Kyttä et al., 2016).

The definition by Bramley et al. (2009, 2010) represents a "people-oriented approach" to social sustainability, emphasizing social cohesion and integrity, social stability, and improvement in the quality of life (Chiu, 2003). In our view, the linking of social sustainability with urban form is essential. We argue that the multifaceted, controversial social outcomes of densely built urban settings reported by earlier studies can partly be because social sustainability has rarely been studied in a context-sensitive manner identifying the specific ways urban structural characteristics contribute to the accessibility of individually meaningful places and services and the potential experiential and wellbeing outcomes.

In the Urban Happiness study, an online PPGIS methodology was used for a place-based study of urban and suburban contexts. Over 3000 respondents located their meaningful places and reported the experiential and wellbeing outcomes. The place-based approach allowed the study of residents' urban networks in an individually sensitive manner and anchored findings on the urban fabric in a contextually sensitive manner (Kyttä et al., 2013, 2016).

The findings indicated that the social sustainability of densely built urban neighbourhoods is a highly complex and context-dependent issue and that densely built urban neighbourhoods can include characteristics that support social sustainability (Kyttä et al., 2016). Residents living in the Helsinki metropolitan area in Finland generally evaluated the quality of environment higher in urban than suburban neighbourhoods in terms of appearance, atmosphere and social and functional quality. Functional quality scored highest in both urban and sub-urban settings, which is in contrast with the study by Walton et al. (2008) in New Zealand, where functional quality generally scored the lowest. Our finding can be related at least partly to the strong functionalistic tradition in Finnish architecture and urban planning.

The analysis of the association between urban density and perceived environmental quality revealed first a curvilinear association, where the perceived quality peaked at a rather high level of urban density, around 100 housing units per hectare (Figure 3.1). Further steps of analysis applying structural equation modelling uncovered significantly differing patterns in the way social sustainability evolved in two contexts, that is, in urban and suburban settings (Kyttä et al., 2016). While urban density promoted easy access to everyday services in both urban and suburban contexts, the experiential outcomes varied. In the urban context, easy access to services contributed to higher perceived environmental quality, whereas in the suburban setting, the closeness of services as well as the increasing density decreased perceived environmental quality. Closeness to services even had an association

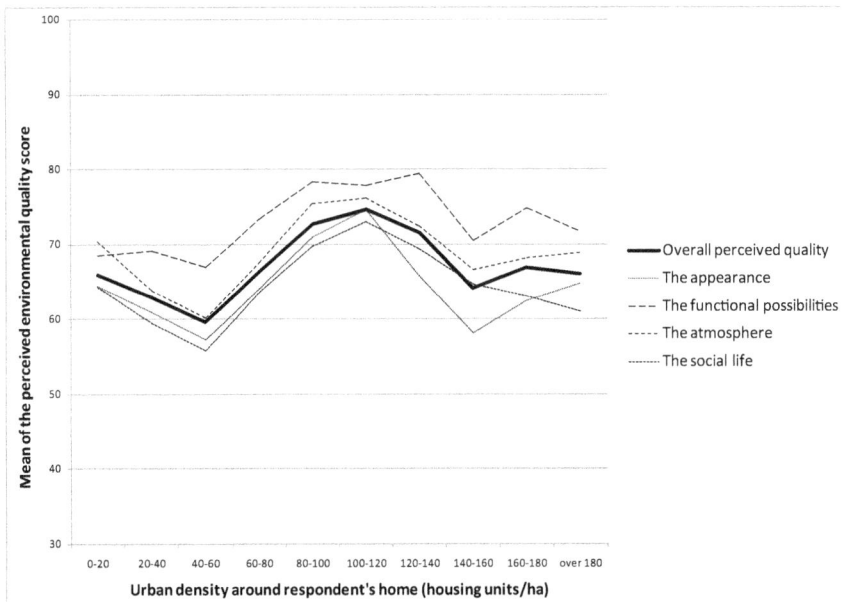

FIGURE 3.1 The co-variation of urban density and perceived environmental quality

with wellbeing, but again these were opposite in the two contexts: in the urban context, closeness to services had positive outcomes, whereas in the suburban context, it had negative outcomes. In both contexts, perceived environmental quality mediated these outcomes. Also in other studies, the closeness of daily services and easy access for errands has been shown to include positive physical health outcomes by promoting active travel modes (Durand et al., 2011).

According to our study in Helsinki, urban density can both meet the ecological goals for efficient urban structure and support social sustainability by providing good everyday service accessibility and the resulting positive experiential and wellbeing outcomes. In the urban context of our study, this happy story held true. In the suburban context, conflict partly existed because the two essential dimensions of social sustainability, accessibility and experiential outcomes, did not support each other. These findings may suggest that only in a setting that is urban enough, do the benefits of service accessibility bloom. This finding can give some support to optimal centrality theory (Cicerchia, 1999; McCrea and Walters, 2012) which states that an optimal level of urban density can be found where access to services and facilities can be guaranteed without overwhelming urban problems such as pollution and traffic congestion. A challenge remains for urban planners on how to improve positive quality place and service accessibility and related experiential outcomes in suburban contexts as well.

The core research question in the Urban Happiness project related to the association between the level of urban density and the perceived urban quality. The finding that a moderate level of density, about 100 housing units/hectare, was perceived most positively by inhabitants, was novel and intriguing. Being able to answer questions like '*What is the association between the various levels of urban density and perceived environmental quality?*' allows the use of PPGIS as a diagnostic tool in research and in participatory planning (Horelli, 2002). While the study generally revealed rather positive experiential outcomes of urban density, the same dataset also exposed inhabitants' strong affection towards green settings. We found that positive places were located more often in green areas than in areas representing other land-use patterns (Kyttä et al., 2013).

In the Urban Happiness project the diagnostic analysis related to the preferred density level was initiated and performed by researchers. Although the findings raised some public debate, they were not explicitly used as a justification for the target level of density applied in the City Plan of Helsinki. In our view, the kind of context-sensitive approach realized in this study could help a planner not only define the acceptable density level, but also help find potential locations for infill projects without high experiential costs (Kyttä et al., 2013). In the next section we describe how the public participatory process of the Helsinki City Plan evolved using the same mapping methodology.

The planning innovation

The innovation in Helsinki was the development of a sustainable city through large-scale public participation. While planners generally strive to densify and retrofit urban areas to improve ecological sustainability, residents are usually critical of the changes in their living environment that may threaten the qualities they value. The wish to have one's surroundings intact forms a challenging starting point for public participation in densification projects. Evidence from the Urban Happiness project reported briefly above was nevertheless able to show that urban density can include clear benefits both in terms of better accessibility as well as experiential and health outcomes (Kyttä et al., 2016). Below we explore how the inhabitants of Helsinki viewed the future of their city and the urge to densify urban structure and the need to protect green areas.

The arranging of public participation has become an elementary part of urban planning practices in Western democracies. Although participation has been justified through democratic rights and procedural justice, the influence of participation on decision-making and actual outcomes has remained contradictory (Bäcklund and Mäntysalo, 2010; Beresford and Hoban, 2005; Irvin and Stansbury, 2004) and received less attention among researchers. In Finland, stronger self-government and increased delegation led to a renewal of the Planning and Building Act in 2000 that demanded a more participatory approach in all planning processes. In practice, realizing participatory planning has not been easy because planners have lacked adequate skills and tools to arrange participation. Also, the success of participation has not been holistically evaluated, leading to continuous suspicion of the benefits of public participation. Understandably, the outputs and outcomes of participation have received less attention because they are not easy to measure (Rowe and Frewer, 2004). Problems related to data availability, long time horizons and the difficulties in measuring social and environmental outcomes are among the challenges included in attempts to evaluate the influence of public participation. The public participation process of Helsinki City Plan provided a possibility to study whether some of these challenges of public participation can be tackled using PPGIS methodology.

Innovation process

In the early stages of Helsinki City Plan process, during autumn 2013, a large-scale public participation effort was arranged by the planning office of the city. An online, place-based dataset was collected that included responses from nearly 4000 residents that marked over 33000 places related to the core themes of the Helsinki City plan (Kahila et al., 2016). The PPGIS survey was developed using the Maptionnaire service by researchers and city planners together. The participants were instructed to consider the future of the city and to locate different places and routes on the maps. The main questions were related to two themes: 1) the possible locations of the new building sites and 2) the important green areas that should be protected. The respondents were able to freely locate these places on the map as well as share ideas for future development. The survey also included questions regarding the participants' background, and questions about the participants' attitudes towards urban development.

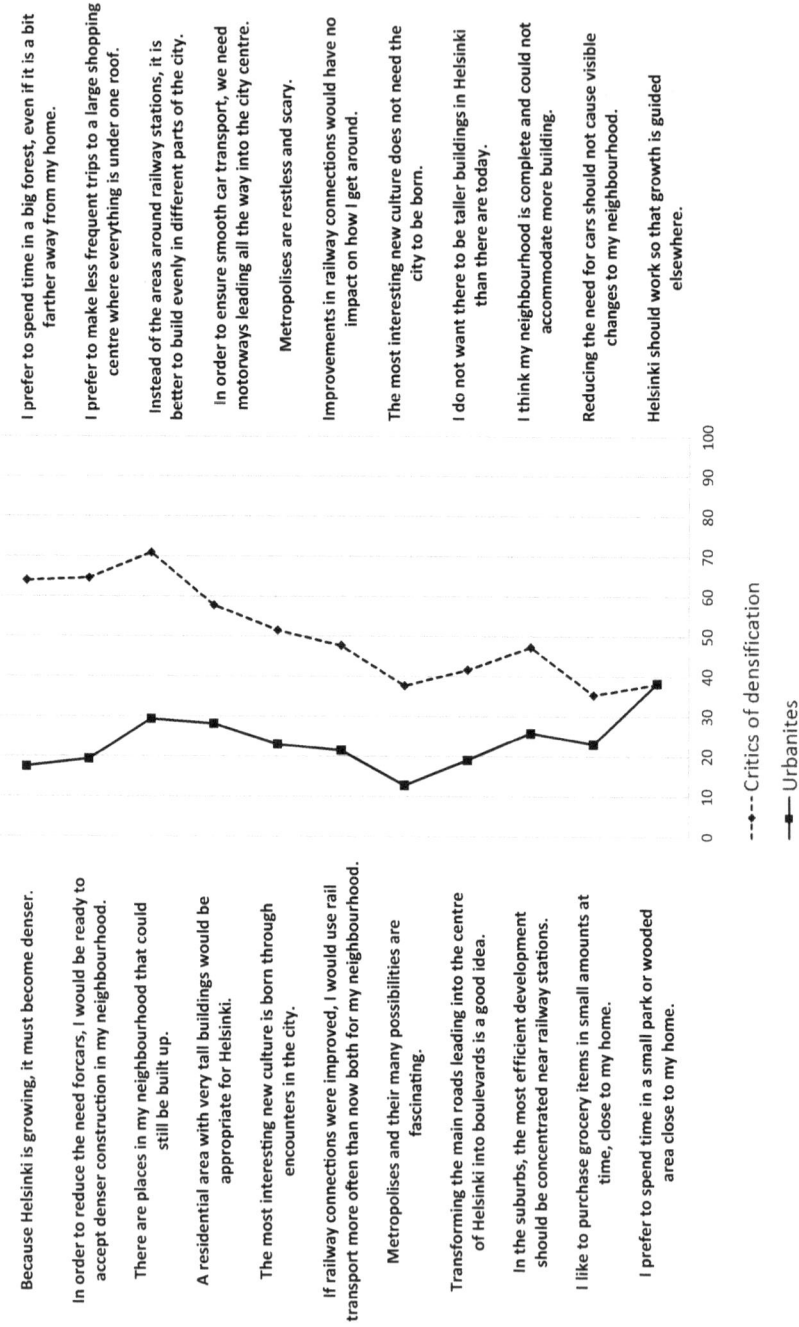

I prefer to spend time in a big forest, even if it is a bit farther away from my home.

I prefer to make less frequent trips to a large shopping centre where everything is under one roof.

Instead of the areas around railway stations, it is better to build evenly in different parts of the city.

In order to ensure smooth car transport, we need motorways leading all the way into the city centre.

Metropolises are restless and scary.

Improvements in railway connections would have no impact on how I get around.

The most interesting new culture does not need the city to be born.

I do not want there to be taller buildings in Helsinki than there are today.

I think my neighbourhood is complete and could not accommodate more building.

Reducing the need for cars should not cause visible changes to my neighbourhood.

Helsinki should work so that growth is guided elsewhere.

Because Helsinki is growing, it must become denser.

In order to reduce the need for cars, I would be ready to accept denser construction in my neighbourhood.

There are places in my neighbourhood that could still be built up.

A residential area with very tall buildings would be appropriate for Helsinki.

The most interesting new culture is born through encounters in the city.

If railway connections were improved, I would use rail transport more often than now both for my neighbourhood.

Metropolises and their many possibilities are fascinating.

Transforming the main roads leading into the centre of Helsinki into boulevards is a good idea.

In the suburbs, the most efficient development should be concentrated near railway stations.

I like to purchase grocery items in small amounts at time, close to my home.

I prefer to spend time in a small park or wooded area close to my home.

0   10   20   30   40   50   60   70   80   90   100

---◆--- Critics of densification

—■— Urbanites

FIGURE 3.2   The profiling of survey respondents to two groups based on their attitudes towards urban densification

The findings revealed that the majority of participants of the survey (61 per cent) had a rather positive attitude towards urbanization of the city of Helsinki and they associated many benefits with densification. The critics of densification (39 per cent) were more worried about losing the qualities that are personally meaningful to them and wished that urban densification could happen elsewhere, not in the immediate vicinity of their homes (Figure 3.2).

The rather positive attitudes towards urban densification were also visible in the mapping by survey respondents of nearly 17,000 potential new building sites on the map. At the same time, the participants also marked about 5000 unique city nature spots that should be protected. These places were located mainly along seaside areas. Figure 3.3 shows the mapping of over 14,000 potential sites for new dwellings. Because many suggestions for new dwellings were located along the main highway corridors leading to the city centre, the mapping results can be interpreted as giving some support to one of the main ideas of Helsinki City Plan. According to the City Plan, about one third of urban infill is planned to be located along highways that will be changed into urban boulevards, allowing the exclusion area to be used for building.[2]

In comparison to earlier studies that have identified strong opposing not-in-my-backyard (NIMBY) groups (Dear, 1992), in Helsinki a supporting YIMBY group was found. This group holds a positive view of retrofitting of the city structure and values the many possibilities of urban lifestyles. The respondents to

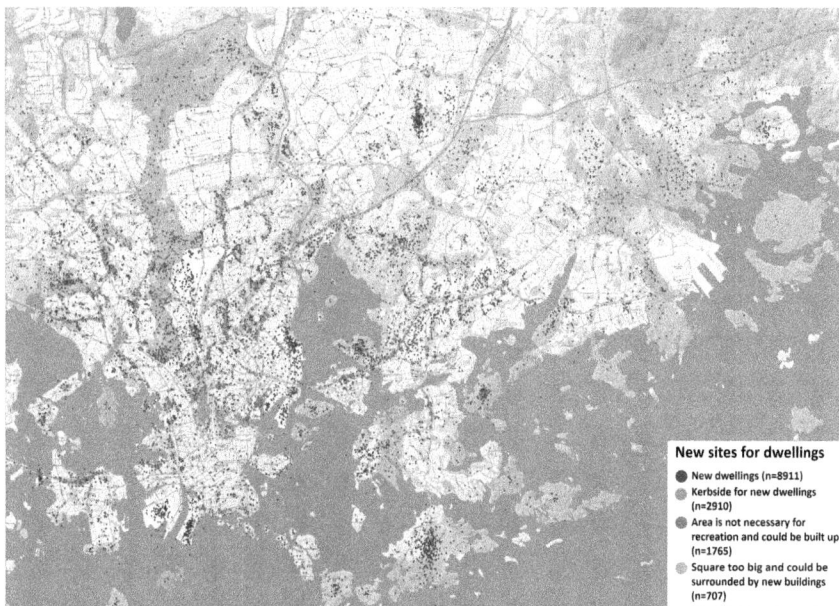

New sites for dwellings

- New dwellings (n=8911)
- Kerbside for new dwellings (n=2910)
- Area is not necessary for recreation and could be built up (n=1765)
- Square too big and could be surrounded by new buildings (n=707)

FIGURE 3.3   The mapping of PPGIS survey participants concerning the locations of potential new residential buildings

the survey were, however, not like-minded when the locations of their suggested new building sites and important green values were compared geographically. Both areas where new building sites dominated and areas where important green values dominated were identified. One example of the latter is the seaside area of Vartiosaari where residents mapped important green values more often than new building sites (Figure 3.4). Although many participants were hoping to increase accessibility to the recreation possibilities of the island that is currently hard to access, they strongly criticized the plans to locate 7000 new inhabitants to the island and to build a tram line through the island.

Also, highly contradictory areas existed that included nearly equal amounts of support for new building sites and green areas to be protected. Some of these places were recognized as areas of conflict even in the local media but have mostly been described as conflicts between residents and urban planners, instead of conflicts between different resident groups. The analysis thus revealed several contradictory places in Helsinki where residents' opinions about future development clashed.

In the City Plan, the city of Helsinki identifies places for future densification. The City Plan proposal and the final plan are composed of 100-metre grid cells that indicate the land-use functionalities and the building efficiency for each cell. We analyzed how the values residents had marked aligned with the densification proposals represented in the proposed and the final plans. Figure 3.5 visualizes this analysis. The extensive PPGIS data collected prior to the release of the Helsinki City Plan proposal and the final plan enabled us to study how the resident input compared to the various versions of the plan.

According to the analysis (Table 3.1), it appeared that a total of 25 per cent out of the important green places that the residents identified were inside the

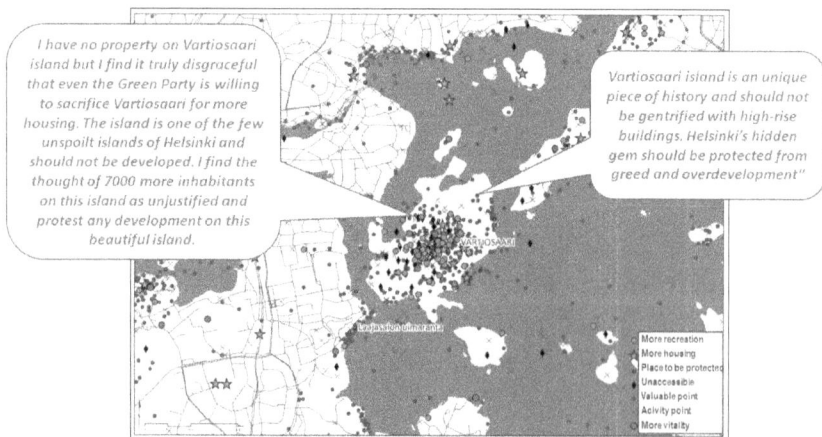

FIGURE 3.4 PPGIS survey respondents' comments about Vartiosaari that attracted many critical comments against densification plans

FIGURE 3.5   Important green values and recommended building sites marked by residents superimposed on the densification areas of the Helsinki 2050 proposal (left) and the final plan (right)

areas proposed for densification in the City Plan proposal, which means that they were threatened by this version of the plan. However, 75 per cent of the important green places were not threatened by the plan proposal. In the final plan, the former percentage had shrunk to 13 per cent. These results indicate an increasing fit between the City Plan and residents' green area preferences, which could be due to the avid public debate and critique that affected the final plan approved in 2016.

During the City Plan proposal phase, the majority (61 per cent) of suggested new building sites were within the densification areas. Because in the final plan the densified area had diminished quite fundamentally (Figure 3.5), the majority of suggested new building sites (62 per cent) now fell outside the densification areas (Table 3.1). This means that the participants of the survey were willing to densify the city of Helsinki even more radically than the decision-makers were ready to do. Thus, the residents suggested many new building sites that extend beyond the areas the city is going to densify. Because residents also marked the majority of green values outside the densifying areas of the plan proposal and even more

TABLE 3.1   Residents' input for important green areas and for new building sites located inside and outside the densification areas defined in the proposed and final Helsinki City Plans

RESIDENT INPUT	Within densification areas (% and number)		Outside densification areas (% and number)	
	Master plan Proposal	Final master plan	Master plan Proposal	Final master plan
Important green areas	**25%** 1515	**13%** 802	**75%** 4476	**87%** 5417
Suggestions for new building sites	**61%** 9580	**38%** 5978	**39%** 6138	**62%** 9856

so in the final plan, urban densification does not severely threaten the green areas that are highly valued by residents. It can thus be concluded that during the planning process the views of the residents and the contents of the City Plan clearly approached each other.

Trade-offs

The Helsinki City Plan process provided an opportunity for a large-scale use of the PPGIS methodology. Compared to the number of residents typically reached via traditional participation methods, the amount of participation input gathered in Helsinki was clearly higher. PPGIS methods are especially valuable in the early phases of the planning process, when detailed background information is needed. Early timing of the participation not only enables the effective use of knowledge, but also supports the acceptability of the plans by fostering trust among participants (Innes and Booher, 2004). In Helsinki, the public participation process could have started even earlier because the City Plan phase was preceded by vision work that outlined the desired land-use solutions for the next decades.

The possibility of comparing user knowledge with the plan proposal and the final plan was a novel opportunity to study the influence of public participation. It is, however, not possible to identify which pieces of knowledge in the long and complex planning process were ultimately influential. The participation methods typically used in planning processes have often relied on discussions between participants and planners. Due to the nature of the information these kinds of processes produce, it is not often possible to analyze the effectiveness of participation on the final planning output. Although PPGIS knowledge is easier to summarize and visualize in cartographic form, further research is needed to analyze how the produced knowledge can be best visualized and analyzed to be used effectively when formulating the planning solutions. Digital tools could also be developed that allow comparisons between PPGIS findings and various alternative plans.

A professor of Urban Planning from Aalto University, Kimmo Lapintie, wrote in his blog in 2014 (http://mahdollisetkaupungit.blogspot.fi/2014/11/) that the Helsinki City Plan documents included few, if any, references to research evidence. Although the city of Helsinki was a project partner of the Urban Happiness study, the findings of the project were not used as a justification, nor any other studies related to social or ecological sustainability. This is very problematic from the point of view of knowledge-informed planning, which highlights the efficient use of diverse, high-quality sources of knowledge to support urban planning. Although planners value the new tools for larger scale participation, more effective use of the produced knowledge should be developed.

Conclusion

This chapter has reported two studies from the city of Helsinki that addressed the challenge common to all cities around the world: how to develop existing environments and plan new communities that are able to combine both social and ecological sustainability. This challenge was topical also in the City Plan process of the city of Helsinki between 2013 and 2016. We reported a real-life participatory urban planning project and preceding research (Urban Happiness project), which both applied an online PPGIS methodology.

The results of the two studies indicated that PPGIS methods are a promising way to collect experiential knowledge from people and a useful addition to the repertoire of traditional participation methods that can help unlock crowd potential (see also Brown, 2015). The findings from the Urban Happiness study suggest that densely built urban environments can promote easy access to everyday services, which is one core criterion of social sustainability. The results related to the other essential dimension of social sustainability, the perceived quality of the environment, were more complex. The perceived quality of the environment peaked at a relatively high level of density, around 100 housing units/hectare. In suburban settings, nevertheless, the closeness of everyday services that urban density entails decreased perceived environmental quality. These results mean that densely built settings can be socially sustainable but investments in the quality of the environment are essential.

In the light of these research findings, the main goal of Helsinki City Plan, the densification of the current, rather dispersed urban structure could potentially promote social, economic and ecological sustainability. The ways how urban densification can promote ecological sustainability are well reported in earlier studies. According to the economic impact assessment of Helsinki City Plan (Helsinki City Planning Bureau, 2015), continuing urbanization and urban densification promote productivity and economic growth through urban agglomeration, among other things. The two studies reported in this chapter suggest that the Helsinki City Plan can also support social sustainability. According to the results of the PPGIS survey, the attitudes of participants were predominantly positive towards urbanization and an abundance of locations for urban infill projects was mapped. When the PPGIS survey data were compared with the plan proposal and the final plan, it appeared that the views of participants were similar to the plan proposal and even closer to the final plan. These findings together with the results of the Urban Happiness study are based on extensive data collection and rigorous analysis, so it can be argued that they represent valid and reliable knowledge that is not always available in planning processes.

Knowledge-informed planning highlights the importance of continuous knowledge construction that should continually feed the planning process through overlapping networks (Davoudi, 2012). Knowledge-informed planning should provide sufficient space to support both instrumental and deliberative notions of planning with an understanding of planning as a practice that should be more open to the broad and often contradictory understandings of people (Kahila-Tani, 2015). So the local experiences of the general public that were collected in the Urban Happiness project could have been used to justify the decisions in the master plan process alongside the rich knowledge produced in the public participation process and by experts and stakeholders. There is no explicit evidence that this happened, so the planning processes in Helsinki could take further steps in the future to develop even more transparent knowledge-informed planning practices.

It is clear that PPGIS tools alone do not make participatory planning better or more influential. In terms of data quality and usability, the localized PPGIS data can provide direct feedback about planning solutions and can be integrated with existing GIS systems. This can help in recognizing user knowledge more equally with other data sets. There is, however, no guarantee that PPGIS data would be more influential than knowledge produced in more traditional public participation processes. Especially when the planning problem is sensitive, a greater level of attention should be placed on data collection strategies to increase PPGIS data reliability and validity. The collected data should also be open to participants and jointly analyzed and debated in a deliberate process.

While many cities share similar challenges to Helsinki with respect to sustainable development and public participation, it is interesting to consider whether the ways the city of Helsinki has used the PPGIS approach would be transferable to other contexts and planning situations. In Finland, the planning sector has welcomed the online PPGIS surveys as a new tool for participation. Most big cities already use PPGIS tools in planning and management. Some Finnish cities, such as Lahti, have made place-based knowledge production an elementary part of their planning practices and have developed ways that various sectors in the city can collect and use the PPGIS data.

The PPGIS approach is also of interest to the planning sector elsewhere. So far Maptionnaire surveys have been used in over 80 countries and they have reached over 400,000 participants, who have made more than 800,000 place markings. The users of the service include the cities of New York and Denver (USA), Stockholm (Sweden), Helsinki, Espoo and Vantaa (Finland) and the Universities of Wisconsin-Milwaukee (USA), Auckland (New Zealand) and Copenhagen (Denmark).

The Maptionnaire surveys used in public participation projects have focused on various planning phases, scales and approaches. The average number of participants in these platforms has been nearly 500, so it can be concluded that PPGIS tools can reach a relatively large number of participants. The planning projects vary in geographical scale, from nationwide surveys to those concerning single buildings. Green- and blue-area planning and management projects with transportation planning projects comprise the majority of cases while statutory master, regional and detailed planning cases are also very common (Kahila-Tani et al., forthcoming).

Participation becomes more effective if it takes place early enough in the planning process. In the Maptionnaire cases, both extremes of the planning process stand out. Early initiation has been the most common part of the process, but the methods have also been applied in the evaluation phase, which has thus far been rather neglected in terms of participation efforts. (Kahila-Tani et al. forthcoming). Maptionnaire projects produce high-quality, versatile data, including the collection of positive feedback typically missing in traditional public participation projects.

Because the place-based data produced can be integrated to existing systems, knowledge from participants can be more equally recognized in parallel to other more formal data sets. This can contribute to influential participation that should be the goal of public participation efforts.

The huge, global challenges to promote socially and ecologically sustainability development through urban planning and development cannot be solved using one approach or methodology. The examples of Helsinki discussed in this chapter can, however, pave the way to novel, innovative, people-centred approaches. Without people in mind, the current challenges cannot be solved.

Notes

1 Because we found the original labelling of the two dimensions by Bramley et al. confusing, we renamed them to accessibility (corresponding to social equity) and experiential outcomes (corresponding to sustainability of community).
2 The realization of this idea was recently challenged by the court that decided that the plan is in contravention of administrative law. The implementation of four out of the planned seven urban boulevards was denied. The city of Helsinki responded by making an appeal to the Supreme Court, but the final decision has not yet been made.

References

Bäcklund, P., and Mäntysalo, R. (2010) "Agonism and institutional ambiguity: Ideas on democracy and the role of participation in the development of planning theory and practice – the case of Finland", *Planning Theory*, 9(4), 333–350.

Beresford, P., and Hoban, M. (2005) *Participation in Anti-poverty and Regeneration Work and Research. Overcoming Barriers and Creating Opportunities.* Joseph Rowntree Foundation, New York. Retrieved January 10, 2015, from www.jrf.org.uk/bookshop/eBooks/1859353738.pdf.

Bramley, G., Dempsey, N., Power, S., Brown, C., and Watkins, D. (2009) "Social sustainability and urban form: Evidence from five British cities", *Environment and Planning A* 41(9), 2125–2142.

Bramley, G., Brown, C., Dempsey, N., Power, S., and Watkins, D. (2010) "Social acceptability", in *Dimensions of the Sustainable City*, Eds M. Jenks, and C. Jones (Springer, London) pp. 105–128.

Brown, G. (2015) "Engaging the wisdom of crowds and public judgment for land use planning using public participation geographic information systems", *Australian Planner* 52(3), 199–209.

Burton, E. (2000) "The compact city: Just or just compact? A preliminary analysis", *Urban Studies* 37(11), 1969–2006.

Chiu, R. L. (2003) "Social sustainability, sustainable development and housing development: the experience of Hong Kong", in *Housing and Social Change: East-West Perspectives*, Eds R. Forrest, and J. Lee (Routledge, New York) pp. 221–239.

Cicerchia, A, (1999) "Measures of optimal centrality: Indicators of city effect and urban overloading", *Social Indicators Research* 46, 276–299.

Davoudi, S. (2012) "The legacy of positivism and the emergence of interpretive tradition in spatial planning", *Regional Studies* 46(4), 429–441.

Dear, M. (1992) "Understanding and overcoming the NIMBY Syndrome", *Journal of the American Planning Association* 58(3), 288–300.

Dempsey, N., Bramley, G., Power, S., and Brown, C. (2009) "The social dimension of sustainable development: Defining urban social sustainability", *Sustainable Development* 19(5), 289–300.

Durand, C. P., Andalib, M., Dunton, G. F., Wolch, J., and Pentz, M. A. (2011) "A systematic review of built environment factors related to physical activity and obesity risk: Implications for smart growth urban planning", *Obesity Reviews* 12, e173–e182.

EEA (2006) *Urban Sprawl in Europe: The ignored challenge* (Office for Official Publications of the European Communities, Luxembourg).

Finnish Land Use and Building Act (2000) Available at: https://www.finlex.fi/fi/laki/kaannokset/1999/en19990132.pdf.

Helsinki City Planning Bureau (2015) Economic Impact Assessment, The City Plan of Helsinki. [Helsingin yleiskaava: Taloudellisten vaikutusten arviointi]. A report by the master plan department of Helsinki City Planning Bureau [Helsingin kaupunkisuunnitteluviraston yleissuunnitteluosaston selvityksiä], 2015:1.

Horelli, L. (2002) "A methodology of participatory planning", in *Handbook of Environmental Psychology*, Eds R. Bechtel and A. Churchman (John Wiley, New York) pp. 607–628.

Innes, J., and Booher, D. E. (2004) "Reframing public participation: Strategies for the 21st century", *Planning Theory & Practice* 5(4), 419–436.

Irvin, R. A., and Stansbury, J. (2004) "Citizen participation in decision making: Is it worth the effort?", *Public Administration Review* 64(1), 55–65.

Jaakola, A., and Lönnqvist, H. (2007) "Kaupunkirakenteen kehityspiirteistä Helsingin seudulla" [The development of urban structure in Helsinki region], *Kvartti* 1, 35–45.

Jabareen, Y. R. (2006) "Sustainable urban forms: Their typologies, models, and concepts", *Journal of Planning Education and Research* 26(1), 38–52.

Jenks, M. (2010) *Dimensions of the Sustainable City* (Springer, London).

Kahila, M. Broberg, A. Kyttä, M., and Tyger, T. (2016) "Let the citizens map: Public participation GIS as a planning support system in Helsinki 2050 master planning process", *Planning Practice and Research* 31(2), 195–214.

Kahila-Tani, M. (2015) *Reshaping the Planning Process Using Local Experiences: Utilising PPGIS in participatory urban planning.* Aalto University publication series, 223. Available at: https://aaltodoc.aalto.fi/handle/123456789/19347

Kahila-Tani, M., Kyttä, M., and Geertman, S. (forthcoming) "Does mapping improve public participation? Exploring the pros and cons of using public participation GIS in urban planning practices", *Landscape and Urban Planning*.

Kasanko, M., Barredo, J. E., Lavalle, C., McCormick, N., Demicheli, L., Sagris, V., and Brezger, A. (2006) "Are European cities becoming dispersed? A comparative analysis of 15 European urban areas", *Landscape and Urban Planning* 77, 111–130.

Kyttä, M., Broberg, A., Tzoulas, T., and Snabb, K. (2013) "Towards contextually sensitive urban densification: Location-based softGIS knowledge revealing perceived residential environmental quality", *Landscape and Urban Planning* 113, 30–46.

Kyttä, M., Broberg, A., Haybatollahi, M., and Schmidt-Thomé, K. (2016) "Urban happiness – Context-sensitive study of the social sustainability of urban settings", *Environment and Planning B* 47, 1–24.

LEED-ND (2009) *LEED 2009 for Neighbourhood Development Rating System*, Congress for the New Urbanism, Natural Resources Defense Council and the US Green Building Council. Available at: http://new.usgbc.org/sites/default/files/LEED%202009%20Rating_ND_10-2012_9c.pdf.

McCrea, R., and Walters, P. (2012) "Impacts of urban consolidation on urban liveability: Comparing an inner and outer suburb in Brisbane, Australia", *Housing, Theory and Society* 29(2), 190–206.

Newman, P. (2006) "The environmental impact of cities", *Environment and Urbanization* 18(2), 275–295.

Rowe, G., and Frewer, L. J. (2004) "Evaluating public-participation exercises: A research agenda", *Science, Technology, & Human Values* 29(4), 512–555.

Turok, I., and Mykhnenko, V. (2007) "The trajectories of European cities, 1960–2005", *Cities* 24(3), 165–182.

Walton, D., Murray, S. J., and Thomas, J. A. (2008) "Relationships between population density and the perceived quality of neighbourhood", *Social Indicators Research* 89(3), 405–420.

VandeWeghe, J. R., and Kennedy, C. (2007) "A spatial analysis of residential greenhouse gas emissions in the Toronto census metropolitan area", *Journal of Industrial Ecology* 11(2), 133–144.

Vuori, P., and Laakso, S. (2017) *Helsingin ja Helsingin seudun väestöennuste 2017–2050* [The Population Forecast for the City of Helsinki and Helsinki Metropolitan area 2017–2050]. City of Helsinki: Statistics 12.

Yang, Y. (2008) "A tale of two cities: Physical form and neighborhood satisfaction in metropolitan Portland and Charlotte", *Journal of the American Planning Association* 74(3), 307–323.

Zhang, Y., and Fung, T. (2013) "A model of conflict resolution in public participation GIS for land-use planning", *Environment and Planning B: Planning and Design* 40, 550–568.

4

MEDELLIN, COLOMBIA

Social urbanism to build human security

Luisa Sotomayor

Introduction

Violence, insecurity, and fear of crime can seriously compromise the social sustainability of a city by threatening life and restricting other rights (UNDP 2013). Furthermore, responses to violence typically exacerbate socio-spatial exclusion and create unequal access to security and justice (Muggah 2014). This chapter examines the role of urban policy and planning tools to reduce violence and socio-spatial exclusion in Medellin. Since the late 1980s, Medellin experienced what Jenny Pearce (2007) describes as chronic violence: an environment of high-intensity violence where conflicts are place-based and persistent over time. Complex power dynamics characterized lethal violence, including drug wars, territorial disputes between leftist militia and rightist paramilitary squads, turf wars among rival criminal gangs, and social violence, such as domestic abuse and petty crime. Violent actors interacted with local communities in a context of socio-spatial exclusion, limited social and spatial mobility, and income inequality. Over time, these factors recreated perverse incentives for low-income youth to join violent structures (Salazar 2010). Insecurity perceptions and fear of crime further restricted residents' mobility and the pursuit of opportunities (Angarita et al. 2008). Multiple forms of violence imbricated and amplified each other.

This chapter examines a policy of social urbanism as a comprehensive approach to urban security adopted in the early 2000s to reduce violence and marginalization. Social urbanism recognizes violence, socio-spatial exclusion, and institutional fragility as endemic and interconnected problems (Fajardo 2007). Recognizing that chronic violence cannot simply be overturned with orthodox policy tools (Moser 2004), Medellin's policymakers experimented with a coordinated strategy involving social investments, area-based planning, civic engagement, violence prevention programs, and strategic policing. Between 1991 and 2015, Medellin experienced almost a 90 per cent reduction in violence rates (Doyle 2018).

While national security policies and the changing power dynamics among Medellin's criminal actors help explain Medellin's dramatic violence reduction (Civico 2012; Moncada 2016; Sotomayor 2017; Doyle 2018), Medellin's social urbanism and the security strategies adopted in the 2000s have had an important impact in building human security outcomes. Specifically, social urbanism has helped improve the quality of life and create conditions for social, economic, and political inclusion. Social urbanism has also been instrumental in rebuilding the authority and legitimacy of state institutions. The chapter notes, however, that current security policies have departed significantly from previous programs aimed to support human security through social development. Indeed, social investments have been increasingly replaced by larger budgets for policing and security technologies. A key challenge for the social sustainability of the city, therefore, will be to continue building equity in a context where a large proportion of the population continues to be socio-economically excluded. Similarly, the city will need to protect advances in all aspects of human security in spite of changes in urban politics.

The city: Medellin

Medellin is the departmental capital and main metropolitan region of Antioquia. It is also the second largest and wealthiest city in Colombia after Bogota, with a population of 2.5 million inhabitants and an estimated 3.9 million for the metropolitan area. The city is embedded in a narrow valley along the Medellin River, called the Aburrá Valley, and expands to the surrounding upland. Medellin has sharp altitude variations from 1,500 to 1,800 metres above sea level, and some low-income settlements in the outskirts reach 2,100 metre hills (Alcaldía de Medellin 2016). Medellin concentrates high densities in geologically challenging areas, some of them vulnerable to environmental disasters. The highest urbanized elevations are in the city's northeastern area, where slopes have typical gradients between 1:4 and 2:5 and households face risks of landslides or flooding (Alcaldía de Medellin 2011). This intricate topography has mediated many of the challenges posed by its uneven urbanization process.

Medellin is administratively composed of six zones, which are divided into 16 urban districts called *comunas*, and five rural townships, called *corregimientos*. Neighbourhoods represent the smallest territorial unit: there are 249 legal neighbourhoods accepted by the Planning Department (by Decree 346 of 2000) and even more settlements not yet recognized as such (Alcaldía de Medellin 2011). While the city's business districts and middle-class residential neighbourhoods are located for the most part in central and southern areas, the urban poor live in self-help neighbourhoods of difficult topographic access in the northern peripheral hillsides of the valley. In terms of infrastructure, Medellin has a network of multi-modal transit including the only metro system (surface and elevated rail) in the country, but access to rapid transit is highly uneven. The Metropolitan Region of Aburrá Valley (Area Metropolitana del Valle de Aburrá – AMVA) is the regional governance institution – a voluntary cooperation agreement – in place to coordinate environmental planning and regional transportation infrastructure projects across ten member municipalities.

Since the adoption of decentralization reforms, in the late 1980s, local governments have been in charge of providing services such as water and sanitation, waste management, urban planning, parks, sports and recreation, mobility, some health services, and primary and secondary education. Although the police and other public security forces are dependent on the National Ministry of Defense, the 1991 Constitution gave mayors and governors some legal responsibilities over citizen security. Mayors are chiefs of police but remain subordinate to the president in decision-making concerning security. Urban security strategies developed at the local level in Colombia have involved ongoing reform and innovations to curb criminality in Bogota, Cali, and Medellin (Ruiz-Vásquez and Páez 2016). The decentralization of some aspects of security has allowed mayors to experiment and integrate *safe cities* strategies, citizenship culture programs, and hotspot policing into their programs of government, and to build local expertise on new conflict management approaches (Gutierrez et al. 2009).

The urban sustainability challenge: Violence and socio-spatial exclusion

Some authors link violence in Medellin to multi-scalar processes of contested state formation, uneven urbanization, and unsolved exclusionary economic dynamics dating back to the 1950s when the city started a rapid urbanization process (Roldán 2003; Hylton 2007; Maclean 2014; Sotomayor 2015b). Throughout the second half of the twentieth century, Medellin received several waves of dispossessed rural migrants escaping violence in the countryside or looking for better economic opportunities. Despite Medellin's prosperous industrial base, the lack of affordable housing confined a growing population to self-help housing in the hillside communes. Housing for these migrants was initially secured illegally through practices of land invasion, which exposed residents to police harassment and evictions by the local authorities (Hylton 2007). For decades, the poor lived in these hillside communes without adequate access to services and infrastructure. With time, self-help settlements consolidated, housing about 50 per cent of Medellin's population (Echeverri and Orsini 2010).

In the 1970s youth gangs and vigilante fronts gained territorial control in peripheral low-income neighbourhoods, spaces where the judiciary system and the police were weak and lacked legitimacy (Ceballos and Cronshaw 2001). In the 1980s and 1990s, violence escalated. This time, violence was associated to an emerging global drug trade, cartel wars, and the involvement of youth assassins for hire called *sicarios*. In the late 1980s and 1990s, *sicarios* were used by drug lords to target police personnel, human rights activists, lawyers, journalists, judges, university professors, or anyone opposing a cartel's leadership (Thoumi 2002). In 1991, for a city that was then 1.6 million, Medellin achieved a notorious record of 6349 violent homicides, or 381 homicides per 100,000 inhabitants, the highest per capita homicide rate in the history of any city. While statistics on violent homicides were reported as a proxy for violence, unreported violence, such as petty crime, extortion, and domestic violence, also affected residents in their everyday life (Angarita et al. 2008).

Medellin was also stricken by the national armed conflict. Although the conflict was more intensely felt in rural areas, between 1994 and 2002 leftist urban militias and paramilitary fronts organized in Medellin. War between these competing groups and their allied criminal gangs intensified as both insurgent structures got involved in drug trafficking. In 2002 there were 650 criminal bands, three paramilitary groups, and four guerrilla troops under the Ejército de Liberación Nacional (ELN), several guerrilla troops of the Fuerzas Armadas Revolucionarias de Colombia (FARC), and a local leftist independent militia with vigilante roots called Comandos Armados del Pueblo (CAP) (Merchán and Arcos 2011). During the 2000s, Bandas Criminales or Bacrim, a new criminal structure with a paramilitary background, emerged to centralize control over a gamut of underground and illicit activities, including micro-traffic of drugs, robbery, and extortion. Violence reduction has thus been underpinned by a volatile mafia peace (Civico 2012; Doyle 2018).

Violence and social sustainability

Cities experiencing chronic violence, like Medellin, are not only shaped by fear and the violent acts themselves which directly compromise the social sustainability of a city, but also by the types of responses that governments, civic organizations, interest groups, and residents adopt and promote in response to perceived threats. Crucially, fear-management responses tend to shape the urban environment in divisive ways that favour private interests over the public good (Muggah 2014). The proliferation of fortified enclaves across Latin America, as "privatized, enclosed and monitored spaces of residence, consumption, leisure, and work" (Caldeira 2000, 114) exemplify hyper-securitized urbanism. These spaces include office parks, gated communities, shopping malls, clubs, and restaurants, often connected through high-speed roads that are intentionally inaccessible to the poor to shield the dominant classes from perceived threats associated with other social groupings. Urban segregation is thus a consequence of crime perceptions, but also, a function of highly unequal power relations among social classes and how those interests play out in the sphere of urban politics (Rodgers 2012).

A similar outcome is the privatization of security. In many Latin American cities, lack of confidence in the state's capacity to provide effective police security has promoted a growing industry of security personnel and technologies for those who can afford them. Insecurity perceptions can equally incite undesirable forms of community problem-solving such as vigilantism, where efforts to take justice in one's own hands are seen as a legitimate approach. Both types of responses to insecurity perceptions are driven by fear and distrust of the police and the judiciary, which further deteriorates the capacity of democratic institutions (Muggah 2014). Often, violence and fear of crime play out in support of political agendas oriented to repressive policies such as *mano dura* or hard-line approaches, which may actually amplify the social gaps and tensions among socio-economic groupings. Low-income young males of colour, living in neighbourhoods perceived as disorderly, tend to be criminalized, stigmatized, and denied rights (Wacquant 2007). Exclusion and repression can block efforts to deter youth from joining criminal structures as violent organizations provide an attractive lifestyle for youth who lose hope for their future (Riaño-Alcalá 2006; Rodgers 2012).

Perceptions of neighbourhood security and personal safety are typically associated with outcomes in other dimensions of community development and social sustainability requiring social interaction, community participation, or concerted action. For instance, community participation or volunteering may be affected by fear of violence. In contrast, feeling safe in one's place is said to enable trust, social cohesion, solidarity, and reciprocity among neighbours. Safety perceptions contribute to build a sense of community, neighbourhood pride, and a sense of place, which mediate the way people relate to their urban environment (Barton et al. 2003; Shaftoe 2000; Dempsey et al. 2011).

Previous urban security policies

Analysts describe the period 1996–2002 as the withdrawal of the municipal state, as local governments gradually cut an important amount of service provision in communes where armed actors held territorial control. Health services and education infrastructure were in many instances abandoned. Police and other governmental authorities lost access to a large portion of these territories (Merchán and Arcos 2011). The upper classes sheltered themselves in a hyper-securitized "city of walls" in the south of the city and in the suburbs. Indeed, until 2003, Medellin's security framework was guided by a belief that urban violence was an outcome of violent dynamics at the national level and that, as a result, there was little for the local government to do, other than to call for the national government to take action. President Cesar Gaviria (1990–1994) adopted a sticks-and-carrots strategy, including, first, repression and law enforcement; second, a policy of peace pacts with leftist militias; and third, a social development approach. This meant warfare against the Medellin Cartel in the period 1989–1993 and repression in marginalized communes, but also small-scale pacts with militias to de-escalate high-intensity conflicts. Finally, Gaviria appointed an advisory council which acted as a taskforce with an autonomous budget to improve service provision and promote community development called the Office of the Presidential Advisor for Medellin and the Metropolitan Area – *Consejería Presidencial para Medellín y su Área Metropolitiana* (CPMAM) (Sotomayor 2015b).

In the long term, high levels of repression in the war against the Medellin Cartel victimized and stigmatized local communities, particularly youth. Peace pacts with militias were self-defeating as they conferred local legitimacy to armed groups while weakening the role of state institutions. CPMAM was a relevant institution that contributed to build some state legitimacy, particularly via the Integrated Neighbourhood Upgrading Program (PRIMED). PRIMED experienced some success because it was highly participatory and responded to urgent needs for engagement, environmental risk mitigation, and housing security (Betancur 2007; Angarita et al. 2008). However, in 2000, after completing only one out of two proposed housing phases, the program lost political support and was discontinued (Echeverri and Orsini 2010).

In 2002, the right-wing government of President Alvaro Uribe (2002–2010) took a hardline approach, conducting several large-scale military interventions in an attempt to oust guerrilla fronts and their alleged collaborators, particularly in Commune 13. These operations inflicted a great deal of violence and stigmatization on local communities (Sánchez el al. 2011; Sotomayor 2017). Furthermore, following a peace accord with paramilitary forces, in 2004 the Uribe government imposed on Medellin the responsibility to integrate 20,000 demobilized paramilitary combatants into civilian life. This deal had limited success as it exacerbated social tensions. Many of the former combatants joined criminal bands following demobilization (Civico 2012). Amnesty International (2005) questioned this controversial demobilization, as in their view, it legalized paramilitary influence in the city. This new distorted peace did lower lethal violence rates but not due to increased security, but because a unified leadership in criminal structures emerged with less lethal but still violent intimidation tactics (Abello-Colak and Pearce, 2015; Doyle 2018).

The planning innovation: Human security via social urbanism

Following successful experiences in Bogota and Cali in the early 2000s, Medellin began to experiment with its own municipal security policy grounded on claims to human security. Broadly put, human security involves the protection of a basic set of rights that enable people to develop individual and collective capacities, fulfil their individual and collective aspirations, and contribute to their communities. Human security builds on notions of human development and factors in threats such as food insecurity, environmental disasters, violent conflicts, and their interrelations. It also considers security perceptions, not only in relation to fear of violence, but also those linked to basic conditions that guarantee sustenance and life itself (Magaña 2018; Cardona-Berrío and Sánchez-Henao 2014).

During the mayoral terms of Sergio Fajardo (2004–2007) and Alonso Salazar (2008–2011), the municipality developed an urban policy called social urbanism. Social urbanism emerged in recognition of the multi-dimensional and interrelated nature of violence, marginalization, and fragile governance in low-income communes where armed actors held territorial control. These neighbourhoods, typically in the peripheral hillsides, were also the poorest, and required significant investments in infrastructure and service delivery. Residents were politically disengaged and used to clientelist practices to access state resources (Moncada 2016). In these spaces, state presence was weak and governmental institutions, including the judiciary and the police, had lost legitimacy (Alcaldía de Medellin 2010).

FIGURE 4.1 PUI in the northeastern area
Source: Author

As a response, the municipal government designed an urban upgrading tool, the Integrated Urban Project (Proyecto Urbano Integrado or PUI). PUIs were devised as an area-based planning instrument to redress the unequal distribution of urban infrastructure (Sotomayor 2015a). As Figure 4.1 illustrates, PUIs are characterized by modern and flashy architecture, and highly visible infrastructures, such as cable-car systems, library-parks, and outdoor escalators. Library-parks are devised as a hybrid indoor/outdoor multi-purpose community space, delivering typical library services, but also community programming and social services. In conceiving these projects, Medellin took references from the Barcelona model of urban regeneration and Bogota's public space and citizenship culture programs from the 1990s. Social urbanism uses the symbolic capacity of architecture to transform local imaginaries about places associated with violent events (Sotomayor 2015a). These infrastructures aim to integrate marginalized districts that were disconnected from the city's urban fabric (Brand and Dávila 2011). Public works were introduced with participatory methods to regain state legitimacy. Participatory planning, and a new Participatory Budgeting Program, for instance, served to establish a new "social contract" and improve state-community relations. Crucially, this area-based planning approach helped the municipality regain territorial control from armed actors through geographically differentiated action plans that were context-specific.

Between 2004 and 2013, about 86 per cent of all municipal expenditures went to social investments (Moncada 2016). Many of these investments were allocated to education, health, early childhood, youth development, and housing improvements. PUIs were successful because they disrupted some of the neighbourhood dynamics on which criminality was rooted, by tackling a gamut of issues from uncertainty around housing tenure, to better and more education programs for youth, to the transformation of local stigmas associated with place-based violent events. Starting in 2007, social urbanism gained international recognition, as there was evidence that it had helped reduce unemployment and increase the quality of life by at least 13 points in areas where PUIs were implemented (MCV 2016). Similarly, as Figure 4.2 illustrates, there was a 32 per cent decrease in homicide rates between 2004 and 2007 (from 58 to 37 per 100,000 inhabitants); and almost a 90 per cent drop from its peak point in 1991 (Doyle 2018).

With the reappearance of armed confrontations in 2008, it became clear that gains in violence reduction were still very fragile and dependent on power relations among criminal gangs. Thus, the city created a new intelligence unit, the Information System for Security and Conviviality. This unit is aimed at systematizing security and violence data and collecting new qualitative data through social research to document non-lethal forms of violence going typically unreported. Data are used to focalize security interventions through territorialized strategies articulated with PUI public works (Merchán and Arcos 2011).

The strategy "Safer Medellin: Together we can" (2008–2011) created comprehensive security plans for each territory articulating justice, policing, human rights guarantors, community partnerships, and preventative programs targeting youth. It is characterized by the concentration of resources on specific problem zones, the territorialization of security operations, and the promotion of citizen participation in security initiatives under the principle of co-responsibility (Alcaldía de Medellín 2010).

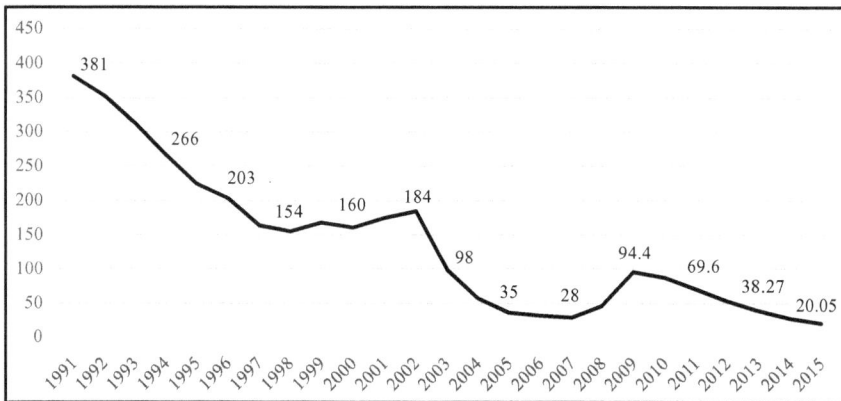

FIGURE 4.2 Homicide rates per 100,000 inhabitants, Medellin 1991–2015

Source: Author's own elaboration with data from Alcaldía de Medellin (2007) and MCV (2016)

According to Merchán and Arcos (2011, 57, translation by author), each action plan considered: "The identification . . . of the problems that affect security and conviviality in each commune in order to generate precise responses according to the nature and intensity of the problems and promote bottom-up governance". Accordingly, differentiated strategies were implemented in areas characterized as *safe zones, safe communities, sensitive places*, and *critical points*:

- Safe zones: strategy for commercial strips and industrial areas, where security problems are typically associated with petty crime and social disorder.
- Safe communities: strategy for residential areas oriented to promote citizenship culture, conviviality, and partnerships to integrate demobilized ex-combatants into civilian life.
- Sensitive places: strategy for territories facing multiple violence dynamics where armed actors hold territorial control.
- Critical points: strategy for areas under dispute by one or more armed groups where open armed confrontations take place (Alcaldía de Medellín 2010).

The initial success of some of Medellin's security policies created a new demand for participation, particularly for communities living in areas where plans for "sensitive places" or "critical points" were implemented. These communities faced a great amount of repression, while better-off communities in areas designated as "safe zones" and "safe communities" collaborated as full and empowered partners (Abello-Colak and Pearce 2015). As a response, networks of social activists, community-based organizations, non-governmental organizations (NGOs), academics and researchers created the *Observatory for Human Security in Medellin*, a partnership formed to further the goals of a human security approach "from below". This new partnership has created instances of community empowerment in the management of their own security (Angarita Cañas and Yepes 2015). The Observatory has documented the first experiences of collective action against armed groups in local neighbourhoods. Reporting has involved a grassroots organizational process, that in turn calls for residents to organize collectively in multiple ways, starting with a discussion of the types of insecurities that affect them, and an ongoing collective reflection on how to humanize the provision of security, in order to guide their advocacy efforts (Angarita Cañas and Yepes 2015, 146). Such state-society complementarity in the promotion of security has been key in advancing citizen-centred understandings and localized actions to promote human security.

Innovation process

National decentralization policy and a shift in urban politics can be credited with enabling policy innovation in Medellin. Decentralization reforms adopted by Colombia in the 1990s created a strong mayoral system and a new structure of opportunity for urban policy to be implemented. In the early 2000s, municipal finances increased substantially, as Medellin was relieved from debt acquired to build the metro system a decade earlier. With a higher budget and new administrative capacities, local government became a highly desirable post for new political forces (Dávila, 2009). In 1999, previously unstructured networks of activists and social organizations came together under a common agenda as part of a new political movement called Citizen Commitment. The movement proposed a redistributive program to improve the quality of life in low-income districts, electing two unconventional mayors consecutively: Sergio Fajardo (university professor with a PhD in Mathematics) and Alonso Salazar (a journalist and social researcher). The rise of Citizen Commitment to power brought a restructuring of local government that activated the transfer of technical capabilities from society (by NGO actors, the private sector, academics, and community leaders with a great deal of expertise) to the local state (Leyva 2010; Sotomayor 2015b).

Once elected, Fajardo and Salazar garnered support from the business elites, who saw an opportunity to lower their security costs and attract foreign investment by supporting social development programs aimed at reducing violence (Gutierrez et al. 2009). To finance social urbanism, Fajardo updated cadastral records and readjusted the property tax. He also improved the management of Medellin's Public Utility Company (Empresas Públicas de Medellín or EPM), which achieved significant surplus profits during the Fajardo-Salazar era. Because social urbanism was supported by an ample cross-class coalition, it did not face much opposition, making possible a radical departure from previous policy approaches (Moncada 2016). The programs of government implemented by these mayors: "Medellin, a Commitment for All Citizens (2004–2007)" and "Medellin: Solidary and Competitive (2008–2011)" were executed expediently, creating a much-needed sense of transformation in local communities. An important implication of this process for polarized cities with acute social class conflicts, such as Medellín, is that redistributive urban reform will be accepted by the local elites if the mayor is able to convey trust by demonstrating managerial skill (Gutierrez et al. 2009).

Trade-offs

The sustainability of social urbanism policies themselves has proved challenging, as they require a large amount of resources and a great deal of coordination, technical capacity, and political will to be maintained. In Colombia, mayors cannot be consecutively re-elected. Mayors Anibal Gaviria (2012–2015) and Federico Gutierrez (2016–2019) have both kept some aspects of social urbanism but departed from its core commitments and principles. While Gaviria and Gutierrez formally claimed their adherence to a citizen-centred human security perspective, in practice both mayors have emphasized punitive securitization measures over comprehensive social development, with larger budgets being increasingly committed to policing and security technologies. A key limitation in this will be to continue building equity and expand infrastructure and services to neighbourhoods that need them. In a city where a large proportion of the population continues to be very poor and excluded, the municipal government and civic society organizations are required to work to improve food, jobs and housing security, climate change resilience, and other responses to security threats affecting marginalized residents.

Social urbanism promoted an important urban reform to improve the equitable provision of security and infrastructure (Sotomayor and Daniere 2018). It also strengthened the institutional capacity of the local state in areas previously off-bounds to the police. However, the area-based approach has been contentious, as it focalizes resources and government interventions in a geographically small terrain, leaving residents outside of the PUI area with an unequal access to valuable infrastructure. Similarly, the differentiated territorialization of policing has meant that many criminal bands that used to have territorial control over PUI zones no longer operate there or have changed their means of violence. In that sense, PUI communities have seen lethal violence decrease but many criminal bands have persisted by changing their violent tactics. Youth gangs in particular have resorted to intimidation, blackmailing, threats and displacement; crimes that often go unreported and enable gangs to maintain their territorial grip. Residents complain that, against police claims, some gangs were not disarticulated but displaced to different neighbourhoods. Thus, new security problems are emerging in communities that were previously safer. Overall, the most complex issue in building lasting security in Medellin will be to establish legitimate and democratic rule of law, given the ongoing criminal domination of the city by gangs with paramilitary backgrounds, and their occasional collusion with state forces (Civico 2012).

Beyond these limitations, PUIs have generated positive local economic development impacts, as all workforce employed to implement the policy was hired from the local neighbourhoods. Similarly, new businesses have flourished along the main avenues and emblematic buildings. Such economic development opportunities have emerged to serve a rising number of local and extra-local visitors who want to ride the Metrocable or visit the libraries for recreation. Education and employment rates have also increased in PUI areas, along with the overall quality of life. For instance, the Quality of Life Survey shows that, between 2010 and 2016, improvements in PUI areas were above the city's average of 1.7 per cent, particularly in Comunas Popular (3.3 percent), Santa Cruz (3.1 percent), and San Javier (2.2 percent) where PUI projects were implemented (MCV 2016, 14).

Conclusion

As the case of Medellin illustrates, violence, insecurity and fear of crime, and their interactions with unsolved structural issues, can seriously compromise the social sustainability of a city. Safety and justice are a precondition for residents of a city to exercise other rights, to move around freely, and participate in all aspects of city life. But violence typically goes hand in hand with other unsolved issues, such as poverty, socio-spatial exclusion, weak democratic institutions, and limited social mobility. These factors can similarly restrict other basic rights. Analysts argue that, when municipal authorities are unable to fulfil their functions and provide human security, cities become fragile or socially unsustainable. Fragile cities are those where institutions lack an adequate level of legitimacy, authority, and capacity to protect core rights and fulfil fundamental roles, such as the protection of lives and property, access to services and infrastructure, and the preservation of basic norms and rights. Thus, the state and its institutions play a fundamental role in enabling all aspects of human security (Muggah 2014).

The transferability of Medellin's policies will depend on the extent to which policy adaptation to new environments is rooted in a contextualized analysis of systemic societal problems and their interaction with local sources of violence. A key lesson is that integrated, multi-sectorial approaches are required to break the inertia of chronic violence and build long-lasting security, including: youth development and violence prevention; reforms to policing and the judiciary to strengthen the rule of law; public investments to promote human development opportunities; improved state-community relationships to legitimize state action; broad and sustained political will to implement bold policy; and finally, effective systems of information anchored in community participation.

References

Abello-Colak, A., and Pearce, J. (2015) Securing the global city? An analysis of the 'Medellín Model' through participatory research. *Conflict, Security & Development*, 15(3), 197–228.

Alcaldía de Medellín (2011) *Medellin en cifras*. Medellín: Departamento Administrativo de Planeación & Observatorio de Políticas Públicas.

Alcaldía de Medellin (2010) *Medellín más segura: Juntos sí podemos. Estrategia de territorialización de seguridad. Política pública de seguridad ciudadana y convivencia*. Secretaría de Gobierno, (https://www.medellin.gov.co/irj/go/km/docs/wpccontent/Sites/Subportal%20del%20Ciudadano/Plan%20de%20Desarrollo/Secciones/Información%20General/Documentos/Seguridad%20y%20Convivencia/Documentos/Medellin%20más%20Segura%202010.pdf), accessed February 19, 2018.

Amnesty International (2005) *The Paramilitaries in Medellín: Demobilization or Legalization?* September, London: Amnesty International.

Angarita, P. E., Gallo, H., and Jiménez, B. (2008) *Dinámicas de guerra y construcción de paz: Estudio interdisciplinario del conflicto armado en la Comuna 13 de Medellín*. Medellín: Universidad de Antioquia.

Angarita Cañas, P. E., and Yepes, C. R. (2015) Alternativas de seguridad de una población víctima de desplazamiento forzado. El caso de la comuna 8 de Medellín. *El Ágora USB*, 15(2), 457–478.

Barton, H., Grant, M., and Guise, R. (2003) *Shaping Neighbourhoods: A Guide for Health, Sustainability and Vitality*. Spon: London.

Betancur, J. J. (2007) Approaches to the regularization of informal settlements: the case of PRIMED in Medellin, Colombia. *Global Urban Development Magazine*, 3(1), 1–15.

Brand, P., and Dávila, J. D. (2011) Mobility innovation at the urban margins: Medellín's Metrocables. *City*, 15(6), 647–661.

Caldeira, T. P. (2000) *City of Walls: Crime, Segregation, and Citizenship in São Paulo*. University of California Press: Berkeley.

Cardona-Berrío, N. A., and Sánchez-Henao, C. (2014). Políticas públicas de seguridad en Medellín: lecturas del problema de in-seguridad desde el enfoque de la seguridad humana.

In *Trans-pasando Fronteras*. Vol. 6, Cali, Colombia: Centro de Estudios Interdisciplinarios, Jurídicos, Sociales y Humanistas (CIES), Facultad de Derecho y Ciencias sociales, Universidad Icesi, 119–138.

Ceballos, R., and Cronshaw, F. (2001) The evolution of armed conflict in Medellín: An analysis of the major actors. *Latin American Perspectives*, 28(1), 110–131.

Civico, A. (2012) "We are illegal, but not illegitimate." Modes of policing in Medellin, Colombia. PoLAR: Political and Legal Anthopology review, 35(1), 77–93.

Dávila, J. D. (2009) Being a mayor: The view from four Colombian cities. *Environment and Urbanization*, 21(1), 37–57.

Dempsey, N., Bramley, G., Power, S., and Brown, C. (2011) The social dimension of sustainable development: Defining urban social sustainability. *Sustainable development*, 19(5), 289–300.

Doyle, C. (2018) "Orthodox" and "alternative" explanations for the reduction of urban violence in Medellín, Colombia. *Urban Research & Practice*, 1–19.

Echeverri, A., and Orsini F. (2010) Informalidad y urbanismo social en Medellín. In Hermelin-Arbaux, M. Echeverri, A., and Giraldo J. (eds), *Medellín: Medio ambiente, urbanismo, sociedad*. Centro de Estudios Urbanos y Ambientales, Universidad Eafit, Medellin, 130–152.

Fajardo, S. (2007) *Del Miedo a la Esperanza*. Alcaldia de Medellín.

Gutierrez, F., Pinto, M. T., Arenas, J. C., Guzman, T., and Gutierrez, M. T. (2009) *Politics and Security in Three Colombian Cities*. Working Paper 44, Cities and Fragile States, DESTIN, LSE, (http://www2.lse.ac.uk/internationalDevelopment/), accessed March 12, 2018.

Hylton, F. (2007) Medellín's makeover. *New Left Review*, 44, 70–90.

Leyva, S. (2010) El proceso de construcción de estatalidad local (1998–2009): La clave para entender el cambio de Medellín? In Hermelin-Arbaux, M., Echeverri, A., and Giraldo, J. (Eds), *Medellín: Medio ambiente, urbanismo, sociedad*. Centro de Estudios Urbanos y Ambientales, Universidad Eafit, Medellin, 271–293.

Maclean, K. (2014) *The 'Medellin Miracle': The Politics of Crisis*. Elites and Coalitions, Development Leadership Program, University of Birmingham: Birmingham.

Magaña, D. M. (2018) El otro paradigma de la seguridad. *Alegatos*, 23(72), 127–150.

MCV (Medellin Cómo Vamos) (2016) *Informe de Indicadores Objetivos sobre Cómo Vamos en Seguridad 2012–2015*, (https://www.medellincomovamos.org/download/informe-de-indicadores-objetivos-sobre-como-vamos-en-seguridad-2012–2015/), accessed May 15, 2018.

Merchán, M., and Arcos, O. (2011) Estrategias de territorialización de la seguridad. Medellín Más Segura: juntos sí podemos. In Alcaldía de Medellín (Ed.), *Laboratorio Medellín*. Catálogo de diez prácticas vivas. Mesa Editores, Medellín 56–77.

Moncada, E. (2016) Urban violence, political economy, and territorial control: Insights from Medellín. *Latin American Research Review*, 51(4), 225–248.

Moser, C. O. (2004) Urban violence and insecurity: an introductory roadmap. *Environment and Urbanization*, 16(2), 3–16.

Muggah, R. (2014) Deconstructing the fragile city: Exploring insecurity, violence and resilience. *Environment and Urbanization*, 26(2), 345–358.

Pearce, J. V. (2007) *Violence, Power and Participation: Building Citizenship in Contexts of Chronic Violence*. Working Paper 274. Institute of Development Studies: Brighton, (https://bradscholars.brad.ac.uk/bitstream/handle/10454/3802/citizenship_chronic_violence.pdf?sequence=1&isAllowed=y), accessed March 18, 2018.

Riaño-Alcalá, P. (2006) *Dwellers of Memory: Youth and Violence in Medellín, Colombia*. Transactions Publishers: New Brunswick and London.

Rodgers, D. (2012) Separate but equal democratization? Participation, politics, and urban segregation in Latin America. In Dennis Rodgers, Jo Beall, and Ravi Kanbur (Eds), *Latin American Urban Development into the 21st Century: Towards a Renewed Perspective on the City*. Palgrave Macmillan: New York, 123–144.

Roldán, M. (2003) Wounded Medellín: Narcotics traffic against a background of industrial decline. In Schneider J., & Susser, I. (eds) *Wounded Cities: Destruction and Reconstruction in a Globalized World*. Berg: Oxford, 129–148.

Ruiz-Vásquez, J. C., and Páez, K. (2016) Balance de estrategias de seguridad para zonas críticas en Bogotá y Medellín (Tema central). Urvio. *Revista Latinoamericana de Estudios de Seguridad*, 19, 53–69.

Salazar, A. (2010) *No nacimos pa' semilla. La cultura de las bandas juveniles en Medellin*. Planeta: Bogotá (4th edition).

Sánchez, Luz Amparo, Marta Inés Villa, and Pilar Riaño-Alcalá (2011) *La huella invisible de la guerra: Desplazamiento forzado en la Comuna 13*. Comisión Nacional de Reparación y Reconciliación/Grupo de Memoria Histórica/Semana/Taurus, Bogotá.

Shaftoe H. (2000) Community safety and actual neighbourhoods. In Barton H. (Ed.), *Sustainable Communities: The Potential for Eco-Neighbourhoods*, Earthscan: London, 230–243.

Sotomayor, L. (2015a) Equitable planning through territories of exception: The contours of Medellin's urban development projects. *International Development Planning Review*, 37(4), 373–397.

Sotomayor, L (2015b) Planning through spaces of exception: socio-spatial inequality, violence and the emergence of social urbanism in Medellin (2004–2011). Unpublished PhD dissertation thesis, University of Toronto, Toronto.

Sotomayor, L. (2017) Dealing with dangerous spaces: The construction of urban policy in Medellín. *Latin American Perspectives*, 44(2), 71–90.

Sotomayor, L. and Daniere (2018) The dilemmas of equity planning in the Global South: a comparative view from Bangkok and Medellín. *Journal of Planning Education and Research*, 38(2), 273–288.

Thoumi, F. E. (2002) Illegal drugs in Colombia: From illegal economic boom to social crisis. *The ANNALS of the American Academy of Political and Social Science*, 582(1), 102–116.

UNDP (United Nations Development Program) (2013) *Regional Human Development Report 2013–2014. Citizen Security with a Human Face*, (http://www.latinamerica.undp.org/content/dam/rblac/docs/Research%20and%20Publications/IDH/Regional_Human_Development_Report_2013-14.pdf) , accessed March 2, 2018.

Wacquant, L. (2007) Territorial stigmatization in the age of advanced marginality. *Thesis Eleven*, 91(1), 66–77.

5

VANCOUVER, CANADA

Equitable sustainability planning
within the post-industrial urban
development model

Wesley Regan and Peter V. Hall

Introduction

The City of Vancouver lies at the heart of a prosperous and growing metropolitan region. Since the 1980s, the City's economy has been based on a real estate-led model of post-industrial development. This model brings with it challenges in terms of equity, social inclusion, cultural sustainability of neighbourhoods, and meeting the needs of vulnerable populations. Still, the City of Vancouver is widely praised for its commitments to sustainability, and it has adopted broad overarching sustainability frameworks with its Greenest City Action Plan (City of Vancouver 2011), Economic Action Strategy (Vancouver Economic Commission 2011), and Healthy City Strategy (City of Vancouver 2014c). In all three strategies the real estate industry looms large, whether as a locus for environmental innovation, as the generator of so-called 'green jobs', or as a partner to support social innovation and provide cultural spaces. The 2011 Greenest City Plan and Economic Strategy together prioritized a decarbonizing but technology-intensive vision of urban development. The 2014 Healthy City Strategy, in contrast, prioritizes issues of equity, affordability, and social inclusion.

In this chapter, we trace how Vancouver's innovative approach to social sustainability unfolded over the past two decades through various overarching sustainability frameworks and their subsequent plans and sub-strategies. We draw on a close reading of these many plans and strategies, as well as our own professional experience and ongoing engaged research on community economic development and social enterprise. Most attention is paid to the efforts of the municipal government in its Downtown Eastside. These efforts have resulted in novel collaborations with communities, non-profits, private industry, and other governments. A watershed moment in this process of planning and policy experimentation and learning was the adoption of the multi-governmental

Vancouver Agreement in 2000. The Agreement, which was implemented over the decade to 2010, left legacies that have informed and been adapted in subsequent equity-based social sustainability planning efforts.

In this trajectory, we identify innovation in *how* sustainability planning is pursued, *who* is involved in the process, and *what* is considered as part of socio-economic sustainability planning.

The city: Vancouver, Canada

The City of Vancouver (2016 population 631,486) is the main urban core of the Metro Vancouver region (2016 population 2,463,431) which is both physically enhanced and constrained by mountains, oceans, rivers, and belts of agricultural and park lands. The colonial settlements of Vancouver are young by global standards. The City only turned 130 years old in 2016. As a centre for colonial commerce, the urban economy has long depended on assembling the financial, service, and human resources for resource extraction, in processing and exporting such resources, and in providing a gateway for the inflows of money, people, and goods that these resources afford. A diverse post-industrial economy of services, amenities, and residential development has been overlaid on the older industrial and resource-based economy (Hutton 1997). In addition to these, the film and television industry, technology, web-based applications, and other creative sectors have increasingly defined the city's economic priorities within the emerging 'knowledge economy' (Vancouver Economic Commission 2018).

The City is one of 23 municipalities that together comprise a weakly federated metropolitan government, known as the Greater Vancouver Regional District. Across Canada, local governments are subordinate to their respective provincial governments. Since the 1990s, they have experienced strain as an increasingly broad and complex set of responsibilities traditionally falling under the work of senior government ministries has been downloaded onto them (Duffy et al. 2014). The constitutionally weak place of local government in Canada means that the City must necessarily operate within a multi-scale governance system to secure resources and resort to entrepreneurial approaches that leverage private sector development. This brings its own set of challenges in terms of equity, the public good, the creation of public risk, and private profit (Tasan-Kok 2010).

As senior Canadian governments turned to more neoliberal policy stances starting in the 1980s and 1990s, local governments came to rely more and more on real estate development to fund the gaps left by their retreat. This has been done through aggressive up-zoning to maximize property values, and with such tools as Development Cost Levies and Community Amenity Contributions collected by the City to fund both essential infrastructure and desired public amenities. Despite the retrenchment of funding for cities from senior governments over the past few decades, critical voices have observed that local government risks becoming 'addicted' to these value-capture tools that are a feature of neoliberal entrepreneurial cities seeking to leverage development as a means to finance growth (Condon 2014; Ladner 2014).

Though experiencing robust growth today, Vancouver was once a place of industrial restructuring and decline, losing population briefly in the mid-1970s. By the mid-1980s, population growth and a return to economic health was demarcated by Expo 86, the 1986 World's Fair 'mega event' and the related processes that invited global real estate investment into the city (Olds 1998; Essex and Chalkley 2004; Aragao and Maennig 2013).

The urban sustainability challenge: Social sustainability in the greenest city

It is in this decisive turn towards a post-industrial real estate-based model of urban growth, which Vancouver shares with other cities in Canada and elsewhere (Olds 2001), that we find the sources of the current urban sustainability challenges and tensions.

The urban sustainability challenge examined in this chapter is equity: *how is social sustainability planning approached in a city that has embraced a post-industrial real estate model of urban development*? This model, which prioritizes technologically intensive innovation and attraction of a highly skilled workforce and foreign capital, is now synonymous with increased costs of living and inequities that are contributing to an 'urban crisis' (Siemiatycki et al. 2016; Florida 2017). In the case of Vancouver, much focus is placed on the city's Downtown Eastside (DTES), a neighbourhood which has undergone dramatic transformation from being the original core of downtown commerce, to carrying the ignoble mantle of being 'Canada's poorest urban postal code' (Matas and Peritz 2008). Here is where different visions of social and economic sustainability collide, creating tension and resistance. It is also where Vancouver's most innovative planning processes and practices to address social sustainability can be found.

Beginning in the late 1960s, the role of Vancouver's DTES changed from providing temporary lodgings and entertainment for shift workers in the resource sector, to being a source of cheap and often inadequate housing for unemployed and low-income residents. In 1997, after years of disinvestment and decline, the City of Vancouver declared a Public Health Emergency as disease and overdose rates from illicit drugs rose sharply (BC Centre for Disease Control 2012). The City created 'A Program of Strategic Actions for the Downtown Eastside', aimed at reducing drug addiction, crime, and street disorder, improve housing conditions, and to 'Help Community People to Find Allies and Seek a Common Future'.[1] This included convening a City Caucus of elected officials with the goal of moving towards a 'Tri-Government and Community Partnership'. This eventually developed into a multi-governmental framework known as the Vancouver Agreement (VA), signed in 2000 between the City, the Province, and the Federal Government. Over 20 ministries and government bodies were involved in the work of the Agreement. The VA also engaged with community agencies, non-profits, and corporate partners like Vancity Credit Union and Bell Canada. However, residents themselves with the lived experience of poverty are not listed among the partners who created and implemented the Agreement.

The Agreement's Economic Revitalization Strategy, adopted in 2004, focused on 'Revitalization Without Displacement'. The cornerstone initiative of the strategy was a large-scale Community Benefits Agreement, which saw CAD$42 million in local procurement from local social enterprise and other businesses, and 124 jobs for DTES residents with employment barriers, created through the Olympic Athletes Village real estate development (Western Economic Diversification Canada 2010, 24).

Following this, between 2011 to 2016, the ruling Vision Vancouver Party, a 'big tent' green and progressive party formed in 2005 and installed into office in 2008 amidst the start of the Great Recession, approved three overarching sustainability strategies for the city. These were the Vancouver Economic Action Strategy (Vancouver Economic Commission 2011), Greenest City Action Plan (City of Vancouver 2011), and Healthy City Strategy (City of Vancouver 2014c), referred to henceforth as the EAS, GCAP, and HCS, respectively. The first two focused on developing a sustainable 'green economy' to support a high quality of life. The third focused on social development and well-being. Together they comprised the planning sustainability triangle (Campbell 1996).

Meanwhile, as these strategies and plans were being created and implemented, Vancouver became increasingly recognized for its extreme housing unaffordability challenge. The city has seen a frenzied pace of construction for several decades (Gordon 2016). Despite insistence by some that housing affordability can be solved

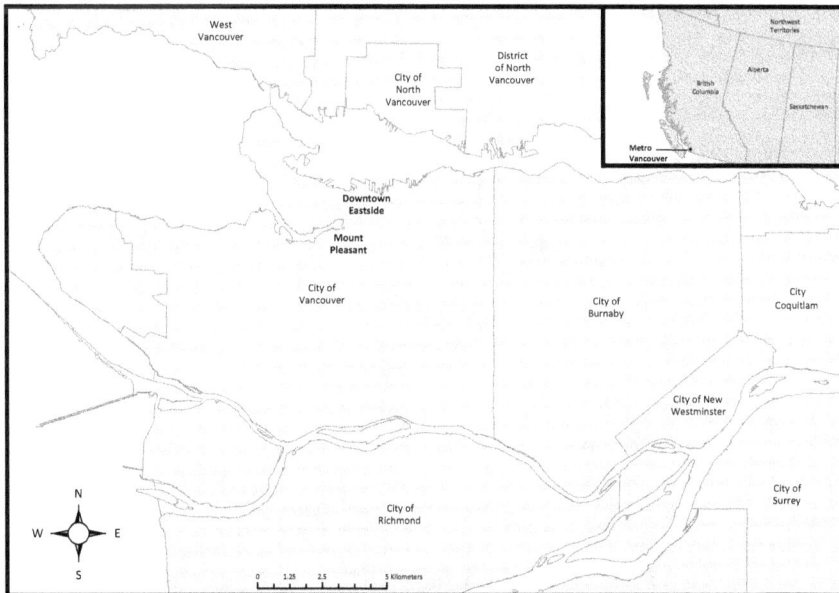

FIGURE 5.1 Downtown Eastside in Vancouver

Source: Department of Geography, Simon Fraser University. Used with permission.

through increasing supply, there have been no corresponding improvements in affordability (Rose 2017). Nevertheless, supply-side arguments to reduce housing prices through development have added pressure on industrial lands, commercial-retail corridors, and lower-income neighbourhoods, including the DTES. These pressures include the encroachment into traditionally more affordable industrial employment lands of non-conforming uses by technology and gentrifying retail activities. Some of these lands are located in the DTES Planning Area and immediately adjacent communities such as Mt Pleasant (Bula 2016; Vikander 2018).

The planning innovation

In response to the growing concerns about displacement, inclusion, and affordability, Vancouver's equity-based planning processes have shown a progressive commitment to participatory planning and co-creative models that include, in particular, residents with lived experiences of poverty, stigmatization, mental health struggles, and other related barriers. These planning initiatives thread forward concepts and values established in the VA, including revitalization without displacement, a preference for collaboration over competition, and multi-sectoral input and governance (Western Economic Diversification Canada 2010, 27–28).

The VA saw different levels of government collaborating with agencies and the private sector to prioritize inclusive economic development. The HCS also abides by a commitment to 'collaborative leadership' and involves close partnership with the provincial health ministry and civil society actors. More inclusive still, the DTES Plan (2014) and the Community Economic Development (CED) Strategy (2016), which both drew on the overarching frame provided by the HCS, are explicitly place-based collaborative approaches, with local agencies, social enterprises and people with lived experiences of poverty forming and co-chairing planning and co-creation committees. In the case of the CED Strategy, they are involved in implementation and monitoring.

The CED Strategy, in particular, offers an example of innovation both in the process to create it and in its expanded definition of the economy. This new economic framework was directly informed by low-income residents who are typically on the receiving end of policy formation and service delivery (the *who*). The Strategy sought not only to consult these local residents, their service providers, social enterprise and small business representatives, but also resourced them through a self-organized co-creation committee with $150,000 over three years (the *how*). This innovation differentiated the CED Strategy from prior types of consultation where resources were used internally by the City or the Vancouver Economic Commission to convene, share research, inform, and consult. Instead the CED Strategy gave residents the resources to do their own research and develop their own processes for engagement and deliberation. This equity-based co-creation process has resulted in a revised vision of the economy in the context of social connection, health, well-being, access, and inclusion that recognizes non-traditional livelihoods, which are often stigmatized or dismissed by mainstream society (the *what*).

The planning innovation we analyze here reveals a shift from externally oriented efforts in firm and talent attraction, to more attention being placed on local and informal exchange. Paradoxically, this shift has entailed more attention in social planning to land and property, including protecting spaces for low-income residents, small business, and social enterprise.

Innovation process

Developed after GCAP and the EAS which focused on environmental and economic sustainability, the HCS is the overarching social sustainability plan for the entire city, described as "a long-term and integrated plan that helps us think, act and work together in new ways to change the conditions that impact the health and well-being of people, places and the planet" (City of Vancouver 2014c, 4). It comprises the 'third pillar' of Vancouver's overall sustainability vision. The HCS involved extensive public engagement in which 10,000 people participated in workshops and idea labs, and online interaction through various platforms. The strategy is built around a 'social determinants of health' framework that draws from international bodies like the World Health Organization, and in a close partnership with the regional health authority, Vancouver Coastal Health. This included the formation of a Leadership Table which brought together the local health authority with non-profits, foundations, private sector leaders, and the City to collaborate on its implementation. This emulated the 'Inner City Caucus' that led to the Vancouver Agreement's multi-governmental model.

The VA, having 'wrapped up' in 2010, had placed significant focus on social sustainability through inclusive economic development. Both the EAS and GCAP entailed a temporary shift in focus away from such equity-based initiatives. The absence of equity and social justice in both GCAP and EAS reflects Schweitzer's (2016) observation that terms like 'conflict', 'justice', and 'equity' are on the whole conspicuously absent from most (North) American sustainability plans. In contrast, when adopted in 2014 as the City's social sustainability plan, the HCS signalled a return to equity-based planning priorities. As such, it picked up the threads of the VA which continued to be woven together in community by civil society champions such as Vancity Credit Union, social enterprises like EMBERS and Potluck Café, and new institutions like the Hastings Crossing Business Improvement Association (HxBIA). Together, such civil society leaders carried forward concepts and practices established in the VA and re-introduced them in municipal processes to influence policy.

The history and role of HxBIA is instructive because it suggests that equitable sustainability planning imperatives can be carried forward by a variety of local governance vectors (see Darchen 2013). HxBIA was formed out of the VA as an initiative supported by Building Opportunities with Business (BOB), one of the lead proponents of the VA's Economic Revitalization Strategy. Incubated within BOB, its founding directors included social housing provider the Portland Hotel Society, and social enterprises Potluck Café and Catering, and United We Can

(City of Vancouver 2010, 3). The new BIA continued to promote the VA focus on revitalization without displacement by adopting a corporate social responsibility mandate that prioritized economic inclusion and social hiring. HxBIA was a partner organization in both the creation of the DTES Local Area Plan, which included the same VA directives, and the DTES CED Strategy (City of Vancouver 2014b, City of Vancouver Community Services 2016). This illustrates how civil society actors continued, through dialogue and experimentation, to improve upon the work of the VA to support inclusive economic development even while the City officially shifted to an environmental sustainability focus that prioritized job creation for the technological elite.

When it came time for the 2016 CED Strategy to be created, those threads of the 2004 strategy – social enterprise, supported employment, inclusive economic revitalization – were brought into a new framework that expanded to include informal economies and stigmatized work (binning, sex work, street vending, etc.). This was communicated through a new framework called the Livelihoods Continuum, which acknowledged the diverse ways in which people made their livelihoods in the DTES' several neighbourhoods (see Figure 5.2). This is a radically different view of the local economy than that put forward by the EAS (Vancouver Economic Commission 2011) being directly informed by people with lived experiences of poverty, stigmatization, and social exclusion through the innovatively designed and resourced co-creation process. It articulated the same inclusive economic revitalization principle introduced in the VA, but from a more resident-informed perspective.

An important starting point that enabled this radical re-think of the economy is the Downtown Eastside Social Impact Assessment (SIA) undertaken in 2013 and published in spring 2014 (City of Vancouver 2014a). This proved to be a foundational tool, created in 2010 in response to growing community concerns about gentrification, and used as a companion document to the DTES Local Area Plan, in order to:

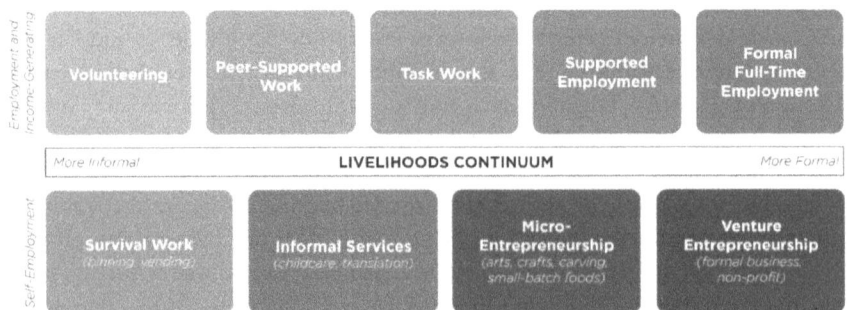

FIGURE 5.2 The Livelihoods Continuum adopted in Vancouver's DTES CED Strategy

Source: City of Vancouver Community Services 2016, 7.

manage community assets, amend regulatory by-laws and policies, manage new business and development and foster Good Neighbour practices in broad community partnership. In moving ahead, the goal will be to maximize beneficial opportunities of development for vulnerable populations and low-income residents and minimize negative impacts which may reduce the quality of life for DTES residents.

(City of Vancouver 2014a, 9)

A planning tool similar to health assessments or economic impact assessments, the SIA is described as a process to "analyse, monitor, and manage intended and unintended social impacts of development" by including communities in identifying concerns in an "anticipatory and proactive way" (City of Vancouver 2014a, 9–10).

The SIA undertook an examination of the physical and social changes in the DTES Planning Area's neighbourhoods, including housing stock, rates of poverty, rates of HIV and Hepatitis C infections, trends in mental health and addiction, child vulnerability, and other indicators. It then shared this information and engaged in asset mapping exercises at workshops to identify where important places and other resources were in community, what value they created for vulnerable residents, and what kinds of gaps could be filled. Many assets identified in the process were intangible in nature and included sense of community, being close to friends and family, not being judged, hope, and being able to freely participate in the informal economy. This would be reinforced a few years later in the DTES CED Strategy through the creation and adoption of the Livelihoods Continuum framework (see Figure 5.2).

Based on community input, the SIA laid out actions to retain valued assets, mitigate negative impacts, and otherwise support vulnerable residents in the DTES through a Social Impact Management Framework that would rely on "transparent and innovative partnerships with local organizations, agencies, groups and individuals" (City of Vancouver 2014a, 56). Along with this a Community Asset Management Program, a Community Based Development Program, Regulatory Tools, and Good Neighbour Practices were recommended to be developed. These were eventually reflected in the inclusionary zoning that rebuffed market condominiums in the Community Based Development Area (CBDA) of the DTES Plan, and in a 'neighbourhood fit tool' – a social impact self-assessment form that all new developments and business permit applicants had to fill out and submit for review to the City's Planning Department.

Drawing heavily from the SIA, the DTES Plan utilized a participatory planning process whereby a Local Area Planning Committee comprised of approximately 40 representatives from a diverse set of stakeholder groups was formed, and which privileged representation from vulnerable and low-income residents. As the Plan states in its introductory chapter: "The overarching-goal of this plan is to make the DTES a more liveable, safe and supportive place for all of its diverse residents, in other words, a healthy neighbourhood for all" (DTES Plan 5). The phrase 'a healthy city for all' deliberately echoes the HCS, which the plan uses, along with the SIA as foundational documents. The DTES CED Strategy also received support when the motion to approve the DTES Plan by Vancouver City Council included additional instruction to staff to create a "coherent community economic strategy for the DTES".[2]

Nested within the established DTES Plan and HCS, the DTES CED Strategy continued to weave together pre-existing directives, principles, and practices. The strategy utilized a public participation vehicle in the form of a co-creation committee that eventually evolved into a strategy implementation body, emulating both the DTES Local Area Planning Committee and Healthy City Leadership Table. This body, called CEDSAC, was resourced by the City with $150,000 to be used over three years as the committee saw fit, in order to engage in the co-creation and co-implementation of the strategy. It worked in close cooperation with the City, but with autonomy. CEDSAC was formed by self-selecting individuals and organizations who responded to a call from the City to serve. The call remained open after the process began in order to include potential committee members who had not received word in time. This helped to ensure that residents living more chaotic lives could freely participate in the committee as a whole or in any of its several working groups – which was reflective of the committee's guiding principles of 'Resident leadership, inclusive economies, reconciliation, respectful co-creation, and systemic change' (CEDSAC 2018).

This resulted in staff presenting a strategy to City Council that abided by the same inclusive economic revitalization principles established in the VA, but with an expanded recognition of opportunities for residents in the context of employment, purposeful activity, and other forms of income generation. This included opportunities that more closely reflected the needs and abilities of low-income residents, which had previously been considered inferior or undesirable.

In this section, we have demonstrated that the planning innovation in Vancouver has not occurred in one single instance but evolved over time, with both government and civil society actors pulling forward and building on concepts, principles, and practices. Increasingly these sustainability innovations have evolved towards more participatory and inclusive planning processes that empower civil society with resources and more equal standing with government. In the process, more radical and unconventional approaches to social sustainability and economic inclusion have now been formally recognized in municipal policy.

Trade-offs

Our following analysis of trade-offs demonstrates how efforts at social sustainability and equity-based planning can be at odds with initiatives of neoliberal entrepreneurial cities that privilege real estate development and corporate firm attraction. In this context, Vancouver's dual visions of economic development and social sustainability in the EAS and CED Strategy are divergent at best, and incompatible at worst. Illustrative of this is the re-purposing of the city's old police station in the DTES into a centre of innovation.

Given the persistent housing affordability challenges in the city, the City's EAS exacerbated tensions in low-income communities when it recommended the City redeploy its real estate assets towards hi-tech development. This was attempted by the strategy's proponent, the Vancouver Economic Commission (VEC), as it sought to re-purpose the City's old police station at the corner of Main Street and Hastings Street in the DTES. Initially proposed was a "central technology hub, or centre of innovation that can offer a complete package of support to entrepreneurs in the early and mid-stages of growth" (Vancouver Economic Commission 2011, 7). The proposal was quickly opposed by residents and eventually re-worked to become a healing and community capacity-building centre and co-working space, more appropriate to the needs of community and more in line with the directives of the equity-based plans and strategies like the HCS, DTES Plan, and CED Strategy (Bula 2013; Smith 2014). Taking cues from the EAS, the CED Strategy similarly proposed that the City invest in social purpose real estate collaborations to free up space for community groups to innovate and implement equity-based economic empowerment initiatives.

In addition to the old police station at 312 Main, renamed the Centre for Economic and Social Innovation, multiple sites have been recognized in which community groups may implement programs and projects (see Figure 5.3). These include a re-purposed City-owned vacant lot that has become the City-sanctioned home of the DTES Street Market, a collaborative of street vendors who had raised concerns about stigmatization and criminalization of their livelihoods, and a

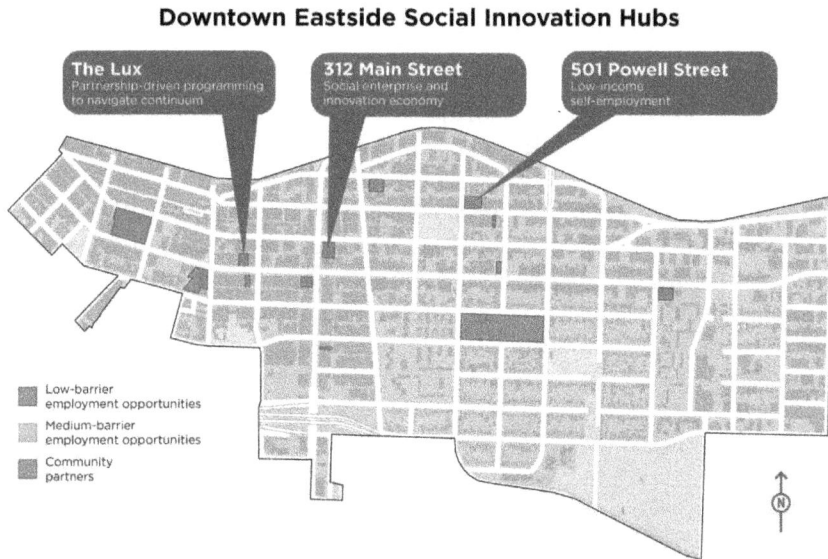

FIGURE 5.3 Location of social innovation hubs in the DTES Planning Area

Source: City of Vancouver Community Services 2016, 9.

City-owned vacant retail unit in the ground-floor of the Lux SRO Hotel. The latter houses the new Eastside Works collaboration led by social enterprise EMBERS and the Urban Core Community Workers Association. This partnership is creating programming and services to help residents with multiple barriers to navigate opportunities within the newly established Livelihoods Continuum economic framework.

This demonstrates how different ideas about *what* constitutes sustainable planning can emerge when publics *who* are traditionally excluded are included, and furthermore, when innovative ideas and non-traditional ways of thinking and doing are included in those plans (the *how*). Who is at the planning table matters, and the process itself that a planner chooses can impact who is able to participate and therefore what kind of ideas inform the plan.

Table 5.1 traces the evolution towards an increasingly inclusive, participatory, citizen-led approach. For example, the co-chairs for the DTES Local Area Planning Process (LAPP) Committee included two non-profits, the Building Community Society, and the Downtown Eastside Neighbourhood Council. CEDSAC worked via a consensus model with the first two co-chairs voted in by members of the committee and subsequent co-chairs similarly voted or acclaimed by consensus.

It took the SIA, the HCS, and the DTES Plan to get there, but the DTES CED Strategy is a culmination of participatory planning processes aimed to ensure that economic policy is increasingly nested in social policy. It seeks to ensure that the needs, desires, and abilities of low-income residents, who are often excluded from formal planning processes, are reflected in municipal policies, programs, and investments.

In some respects, this puts the equity-based community economic development work of the City at odds with the work of bodies like the Vancouver Economic Commission, the lead proponent of the EAS. In addition, the lack of an equity and justice-related focus implies that the GCAP stands more closely with the EAS. The GCAP seeks to foster green growth that is not necessarily more just or inclusive. In many ways, the achievement of the GCAP to reconcile environmental and economic imperatives illustrates the central tensions and dilemmas at the heart of Campbell's (1996) insights, and Connelly's (2007) elaboration of them. It is easier to move in the direction of compromise between two legs of the sustainability stool than to balance all three simultaneously. The HCS and subsequent policy like the DTES Community Economic Development Strategy attempted a rebalancing.

As one of Canada's largest cities, Vancouver's local government is not monolithic. Different departments and arms-length agencies, like the Vancouver Economic Commission, have demonstrated inclinations towards different dimensions of sustainability. Local government faces challenges with divergent values and processes across departments, each with their own mandates, cultures, community stakeholders, partners, and processes for engaging with them. A challenge for Vancouver in the years ahead will be how to reconcile the tensions and contradictions from different visions of economic, social, and environmental sustainability across and within government departments and agencies.

TABLE 5.1 Evolution of participatory planning models in the City of Vancouver

Plan or strategy name	Economic Action Strategy, 2011	Greenest City Action Plan, 2011	Social Impact Assessment Tool, 2013	Healthy City Strategy, 2014	DTES Plan, 2014	Grandview Woodland Plan, 2012–16	DTES CED Strategy, 2016
Citizen participation methods	Selected meetings and interviews with stakeholders	Public engagement Workshops Open houses Online surveys Selected meetings and interviews with stakeholders	Public engagement Asset mapping Interviews with stakeholders Workshops	Public engagement Online surveys Open houses Healthy City Leadership Table	Public engagement Workshops and open houses DTES Local Area Planning Committee	Public engagement Open houses Online surveys Citizens Assembly	Citizen-convened committee Citizen-led working groups Open houses
Model	Consultation	Consultation Limited Participatory	Consultation Participatory	Consultation Participatory	Participatory Limited Co-creative	Participatory Limited Co-creative	Participatory Co-creative
Participatory co-creative body?	No	Internal and External GCAP Advisory Committees	No	Healthy City Leadership Table	DTES Local Area Planning Committee	Grandview Woodland Assembly	Community Economic Development Strategic Action Committee
Participation selection model	Invitation	Open, public engagement online surveys and invitation External advisory	Open	Open (Public input) and invitation (HCS Table)	Open (Public engagement) and invitation (LAP Committee)	Self-selected and randomized	Self-selected non-randomized
Chair or convener	N/A	COV	N/A	Co-chaired by COV and VCH	Co-chaired by community	Third party	CEDSAC
Selection model	N/A	Invitation (Advisory Committees) and Open (public input)	N/A	Invitation	Invitation by co-chairs	N/A	Self-selected
Ongoing or periodic	N/A	N/A	N/A	Ongoing	Periodic	N/A	Ongoing

Source: authors.

These tensions come together most immediately with regards to housing. Hanging over all sustainability questions in Vancouver – and indeed over all current public discourse – is the crisis of housing unaffordability. Housing construction is a driver of jobs in the local economy just as it is a driver of carbon emissions, and as it becomes increasingly unaffordable for those who make their living in the local economy, it potentially harms future job creation, impedes firm attraction, and leads to a range of labour shortages as workers are priced out of communities (Little 2018). The importance of addressing housing affordability in fostering a sustainable economy was even noted in the City's unsuccessful bid to attract tech giant Amazon to locate its second head office there (Bula 2018). Recognizing that adding housing supply alone has not addressed affordability, the City has introduced a recent strategy, Housing Vancouver, which seeks to add more supply but with particular attention to needs and demands along a spectrum of supported housing, rental housing, and lower-end market ownership (City of Vancouver 2017).

Conclusion

Innovations in sustainability planning are not necessarily discrete moments in time, a single plan, site activation, or development. Innovation also unfolds over time as core concepts and ideas travel from plan to plan and strategy to strategy. This is evidenced in Vancouver in the evolution of participatory planning models to empower stigmatized and marginalized groups. In this instance, progress towards a more inclusive process of co-creation has resulted in an innovative take on social sustainability planning. The definition of inclusive economic revitalization advanced in the VA in the 2000s was dominated by the perspectives of the multiple levels of collaborating government. As such, it sought to increase low-income resident participation in the formal economy. After several years of planning exercises to map community assets identified by low-income residents, to understand the aspirations and concerns of vulnerable communities, and increasingly to cede more autonomy and policymaking power to those groups, a new definition of economic inclusion was advanced. No longer about including low-income residents in the formal economy, it is instead about respecting and including informal employment and non-traditional livelihoods as an integral part of the overall economy of the neighbourhood.

How can we learn from these participatory planning innovations in Vancouver so that future environmental and economic sustainability plans in other places can better incorporate justice and equity? Making resources available to level the balance of power in the planning process is an innovation observed in Vancouver's CED Strategy co-creation process. The self-selected advisory committee was provided with financial and staff resources to be used as it saw fit. This recalibrated the power balance of a typical planning process, ensuring that committee members could direct their own 'staff' to further examine issues, seek a second opinion on a matter, organize the administrative needs of the committee and assist in convening and facilitating. This helped to ensure that the content of the strategy was directly informed and prioritized by this association of 30 plus community organizations and individuals with lived experience, and that ownership of its implementation was shared by both the City and CEDSAC (which eventually adopted the name Exchange: Inner-City).

Another consideration is to place less focus on innovation as a thing happening exclusively in one plan or strategy or process and instead consider how these can lay a foundation for or contribute to the conditions in which innovation can arise in the future. Concepts and practices employed in the VA in 2000 and 2004 were not necessarily supplanted by new ideas in the CED Strategy in 2016. Rather these ideas built on commitments to inclusion and multi-sector partnerships, in particular the commitment to include and resource people with lived experience in these processes. It was the continued transference of inclusive economic revitalization as a concept, and participatory collaborative processes as a practice within the organizational culture of the City, from the 2000 VA through the SIA, HCS and DTES Plan, that resulted in the adoption of the innovative Livelihoods Continuum and the investment into social innovation hubs to drive implementation of the DTES CED Strategy in 2016.

While these principles have remained the same, how they have been defined has changed dramatically as people with lived experience of poverty have become more and more involved in policy creation. Through these equity-based social sustainability efforts, they have continued to emulate and adapt the principles, practices, and goals of the Vancouver Agreement but with a radically different perspective on them. This demonstrates that 'new' is not necessarily the same as innovative. Continuing to build on a good idea while trying to do it better is innovation too.

In conclusion, it is important to note that this is not the first example of participatory, citizen-informed planning and policy creation in Vancouver, rather it is part of a legacy of practice (Brunet-Jailly 2008). As Punter (2003) demonstrates, since the 1970s the City has embarked on a range of participatory processes with residents and civil society actors regarding comprehensive master plans and mega projects. Over time it has exchanged practices and ideas with its American counterparts, in particular Seattle, Portland, and San Francisco, in a way that sees it more in tune with American innovations in planning than other Canadian cities. Conversely, Vancouver has also contributed to modern urban design via the tower and podium architecture style of 'Vancouverism' exported abroad, and to public policy leadership on harm reduction via the 'Four Pillars Plan' and its support of overdose prevention sites, a concept now growing in support and expanding to other North American cities. If other cities are to emulate Vancouver's approach to equity-based strategies they too might choose to lay a foundation on which to build capacity, establish and reaffirm that commitment to similar values and practices over time. While such an effort may originate with the innovative ideas of political leaders in a moment in time, Vancouver's example illustrates that those things are *sustained* in the bureaucracy and civil society over the long run.

Notes

1 http://council.vancouver.ca/980728/rr2b.htm
2 http://council.vancouver.ca/20161130/documents/pspc2.pdf

References

Aragao, T., and Maennig, W. (2013) *Mega Sporting Events, Real Estate, and Urban Social Economics – the Case of Brazil 2014/2016* (https://www.econstor.eu/bitstream/10419/92963/1/767777417.pdf) accessed 20 May 2018

BC Centre for Disease Control (2012) *The History of Harm Reduction in British Columbia* (www.bccdc.ca/resource-gallery/Documents/Educational%20Materials/Epid/Other/UpdatedBCHarmReductionDocumentAug2012JAB_final.pdf) accessed 20 May 2018

Brunet-Jailly E. (2008) 'Vancouver: The sustainable city'. *Journal of Urban Affairs*, 30(4), 375–388.

Bula, F (2013) 'High-tech plans for old Vancouver police station divide community', *The Globe and Mail*, 20 March 2013 (https://www.theglobeandmail.com/news/british-columbia/high-tech-plans-for-old-vancouver-police-station-divide-community/article9979861/) accessed 20 May 2018

Bula, F. (2016) 'Gentrification of Vancouver's Mt. Pleasant generating a fight', *The Globe and Mail*, 1 April 2016 (https://www.theglobeandmail.com/news/british-columbia/gentrification-of-vancouvers-mount-pleasant-generating-a-fight/article29504390/) accessed 20 May 2018

Bula, F. (2018) 'Vancouver mentioned its plans to address housing crisis in bid for Amazon HQ', *The Globe and Mail*, 19 February, 2018 (https://www.theglobeandmail.com/news/british-columbia/vancouver-mentioned-its-plans-to-address-housing-crisis-in-bid-for-amazon-hq2/article38025903/) accessed 5 June 2018

Campbell, S. (1996) 'Green cities, growing cities, just cities? Urban planning and the contradictions of sustainable development'. *Journal of the American Planning Association*, 62(3), 296–312

CEDSAC (Community Economic Development Strategic Action Committee) (2018) Mission and Values (https://www.exchangeced.com/mission_values) accessed 20 May 2018

City of Vancouver (2010) *Administrative Report, November 30th* (http://council.vancouver.ca/20101130/documents/a4.pdf) accessed 20 May 2018

City of Vancouver (2011) *Greenest City Action Plan* (http://vancouver.ca/files/cov/Greenest-city-action-plan.pdf) accessed 20 May 2018

City of Vancouver (2014a) *DTES Social Impact Assessment* (http://vancouver.ca/files/cov/DTES-social-impact-assessment.pdf) accessed 20 May 2018

City of Vancouver (2014b) *Downtown Eastside Local Area Plan* (http://vancouver.ca/home-property-development/dtes-local-area-plan.aspx) accessed 20 May 2018

City of Vancouver (2014c) *Healthy City Strategy* (http://vancouver.ca/people-programs/healthy-city-strategy.aspx and http://council.vancouver.ca/20141029/documents/ptec1_appendix_a_final.pdf) accessed 20 May 2018

City of Vancouver (2017) *Housing Vancouver* (http://council.vancouver.ca/20171128/documents/rr1appendixa.pdf) accessed 20 May 2018

City of Vancouver Community Services (2016) *Downtown Eastside Community Economic Development Strategy* (http://council.vancouver.ca/20161130/documents/pspc2.pdf) accessed 20 May 2018

Condon, P. (2014) 'Vancouver's "spot zoning" is corrupting its soul', *The Tyee*, July 14

Connelly, S. (2007) 'Mapping sustainable development as a contested concept', *Local Environment*, 12(3), 259–278

Darchen, S. (2013) 'The regeneration process of entertainment zones and the business improvement area model: A comparison between Toronto and Vancouver', *Planning Practice & Research*, 28(4), 420–439

Duffy, R., Royer, G., and Beresford, C. (2014) *Who's Picking up the Tab? Federal and Provincial Downloading Onto Local Governments*, Centre for Civic Governance, Columbia Institute (www.civicgovernance.ca/wordpress/wp-content/uploads/2014/09/Whos-Picking-Up-the-Tab-FULL-REPORT.pdf) Accessed 20 May 2018

Essex, S., and Chalkley, B., (2004) 'Mega-sporting events in urban and regional policy: a history of the Winter Olympics', *Planning Perspectives*, 19(2), 201–204.

Florida, R. (2017) *The New Urban Crisis: How Our Cities Are Increasing Inequality, Deepening Segregation, and Failing the Middle Class and What We Can Do About It*, New York, Basic Books

Gordon, J. (2016) *Vancouver's Housing Affordability Crisis: Causes, Consequences and Solutions*, Centre for Public Policy Research, Simon Fraser University (www.sfu.ca/content/dam/sfu/mpp/pdfs/Vancouver%27s%20Housing%20Affordability%20Crisis%20Report%20 2016%20Final%20Version.pdf) accessed 20 May 2018

Hutton, T.A. (1997) 'The Innisian core-periphery revisited: Vancouver's changing relationships with British Columbia's staple economy', *BC Studies: The British Columbian Quarterly*, 113, 69–100.

Ladner, P. (2014). 'Vancouver city's CAC revenue stream a debilitating addiction', *Business In Vancouver*, July 28.

Little, S. (2018) 'If people can't afford to work in Vancouver, what happens to the city?', *Global News*, 8 November (https://globalnews.ca/news/3851268/if-people-cant-afford-to-work-in-vancouver-what-happens-to-the-city/) accessed June 5 2018

Matas, R., and Peritz, I. (2008). 'Canada's poorest postal code in for an Olympic cleanup?', *The Globe and Mail*, August 15

Olds, K. (1998) 'Urban mega-events, evictions, and housing rights: The Canadian case', *Current Issues in Tourism*, 1(1), 2–46

Olds, K. (2001) *Globalization and Urban Change: Capital, Culture and Pacific Rim Mega-Projects*, Oxford, OUP Press

Punter, J. (2003). *The Vancouver Achievement: Planning and Urban Design*, Vancouver, UBC Press

Rose, J. (2017) *The Housing Supply Myth*, Kwantlen Polytechnic University (www.kpu.ca/sites/default/files/The%20Housing%20Supply%20Myth%20Report%20John%20Rose. pdf) accessed 20 May 2018

Schweitzer, L. (2016) 'Tracing the justice conversation after "Green Cities, Growing Cities"', *Journal of American Planning Association*, 62(3), 374–379

Siemiatycki, E., Barnes, T.J., and Hutton, T.A. (2016) 'Trouble in paradise: Vancouver's second life in the new economy', *Urban Geography*, 37, 183–201.

Smith, C. (2014) 'City of Vancouver rejects proposed social innovation hub at former police station site', *The Georgia Straight*, February 19

Statistics Canada (2018) *Population Census 2016*, Community Profiles, Statistics Canada

Tasan-Kok, T. (2010) 'Entrepreneurial governance: Challenges of large-scale property-led urban regeneration projects', *Journal of Economic and Social Geography (TESG)*, 101(2), 126–149

Vancouver Agreement (2004) *Economic Revitalization Plan* (www.vancouveragreement.ca/wp-content/uploads/041100_dtes-workplan.pdf) Accessed 20 May 2018

Vancouver Economic Commission (2011) *Economic Action Strategy* (http://vancouver.ca/files/cov/vancouver-economic-action-strategy.pdf) accessed 20 May 2018

Vancouver Economic Commission (2018) *Our Focus* (www.vancouvereconomic.com/focus/) accessed 20 May 2018

Vikander, T. (2018) 'Chipping away at the arts, Vancouver's Red Gate arts space gets the boot', *Metro*, February 13

Western Economic Diversification Canada (2010) *Evaluation of the Vancouver Agreement* (http://publications.gc.ca/collections/collection_2014/deo-wd/Iu92-4-34-2010-eng.pdf) accessed 20 May 2018

6

VAUBAN AND RIESELFELD, FREIBURG, GERMANY

Innovation in the implementation process

Thorsten Schuetze

Introduction

This chapter discusses the sustainable urban development of the German city of Freiburg. The focus is on the districts of Rieselfeld and Vauban, which became good practice examples for strong sustainable urban development. Since the early sustainability movements in the 1970s, Freiburg has developed as the green capital of Germany. A broad range of ecologically motivated activities initiated by residents, stakeholders, and the city government resulted in exemplary sustainable urban development projects. Freiburg is an outstandingly attractive German city due to its special geographical location, the climate conditions, and particularly the mixed population and culture, which are related to the ecological consciousness and self-organization of local citizens. Climate protection and sustainability topics have a high priority for citizens of all age groups (Fraunhofer Institut für Arbeitswirtschaft und Organisation IAO 2013).

In the search for environmentally friendly energy alternatives to nuclear power, the city became the German centre for research, development, and application of solar energy systems. The Ecological Institute was founded in 1977 when the first solar apartment building was also built. In 1981 Europe's first non-university, and today largest, institute for solar energy systems was founded here (Fraunhofer Institute for Solar Energy Systems ISE 2018). In 1986, Freiburg hosted the German State Garden Show, which created among other things the construction and establishment framework for the environmental protection centre "Eco Station". This ecological education centre is one of the oldest in Germany and hosts approximately 600 events per annum and receives 15,000 visitors (Ökostation Freiburg 2018). In the same year, the municipal council resolved to phase out nuclear energy. In 1991, a sustainable waste management and returnable container dictate was introduced (Freiburg Wirtschaft Touristik und Messe GmbH & Co. KG 2016). Moreover, in 1991 the European secretariat headquarters of ICLEI – Local Governments for

Sustainability – were established in Freiburg (ICLEI European Secretariat 1998). Socio-civic initiatives contributed further to bottom-up sustainable development processes, among others supported by the local Agenda 21 office (Freiburger Agenda 21 2018).

In 1992, a low-energy building standard for municipal buildings was introduced and the first energy self-sustaining building was constructed. In 1994 the world's first net plus energy house was built, and the regional public transport network of Freiburg was established. The development of the sustainable districts of Rieselfeld, from 1991, and Vauban, from 1993, became good practice examples on national and international levels for sustainable urban developments within growing cities. For example, the first German multi-family Passive House was built in Vauban in 1999, when the regional energy agency of Freiburg was also founded. In 1996 Freiburg promised to reduce CO_2 emissions by 25 per cent up to 2010 and signed the Aalborg Charta (ICLEI – Local Governments for Sustainability – European Secretariat 2018). The CO_2 emissions reduction goals were updated in 2007 to a 40 per cent reduction up to 2030, and, in 2014, to a 50 per cent reduction up to 2030, and a 100 per cent reduction (climate neutral) by 2050. The setting of associated sustainability targets by the Freiburg Sustainability Council, and their adoption by Freiburg Municipal Council in 2009, established the basis for all related political actions (Freiburg Wirtschaft Touristik und Messe GmbH & Co. KG 2016).

Freiburg was the first city to receive the newly established German Sustainability Award in 2012 for being the most sustainable German metropolis. It was also awarded for the implemented sustainability projects, the city's outstanding engagement in achieving the sustainable development goals, and in particular its unique sustainability infrastructure. A sustainability management unit is associated closely with the lord mayor, and urban societies, offices, agencies, and a forceful body of citizens are committed to the active achievement of sustainable development goals. Environmental and sustainability policy, climate protection and solar technologies became the drivers for sustainable economic, political, and urban developments, and strong unique identifiers for the city and its citizens (Freiburg Wirtschaft Touristik und Messe GmbH & Co. KG 2016).

The City: Freiburg

The City of Freiburg is located in the southwestern part of Germany (Figure 6.1), in the Federal State of Baden-Württemberg.

The city has 28 districts that are subdivided in 48 sub-districts. The city centre, the main important retail centre and reference point for both citizens and tourists, has been closed for car traffic since the 1970s. The city developed a specific market and centre concept that resulted in city districts with their own infrastructures for supply of goods (such as groceries) and amenities, childcare, and education. The city's mobility is based on a tight mass transit system network with trams and buses, and a comprehensive bicycle track network that has been continuously extended

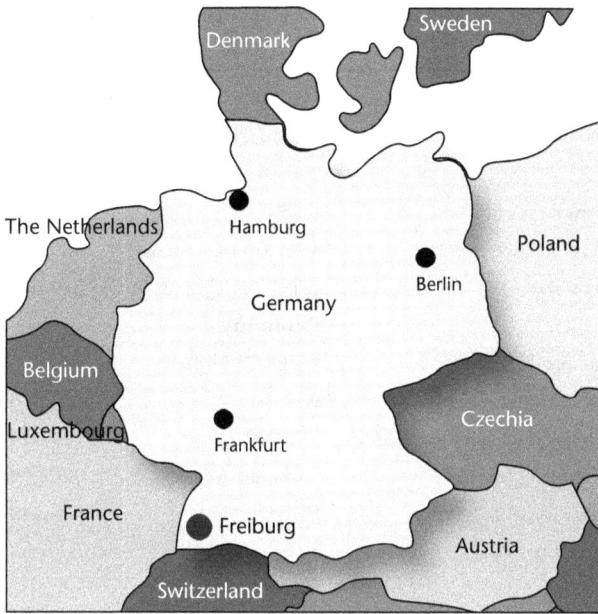

FIGURE 6.1 Location of Freiburg in southwestern Germany

Source: Adapted from Freiburg Wirtschaft Touristik und Messe GmbH & Co. KG, 2016, p. 28

since the 1990s. Freiburg is a tourist magnet with constantly rising visitor numbers. Its attractiveness is related to its nature-connected location along the border with the Black Forest.

A selection of statistical data of general economic and environmental indicators for Freiburg and Germany is summarized in Table 6.1.

Freiburg is regarded as the ecological capital of Germany. According to the federal constitution, the German federal states have legislative authority (particularly educational and cultural sovereignty) except for topics that are the exclusive responsibility of the federal level, such as foreign affairs and defence. In Baden-Württemberg, the Ministry for Economy, Work, and Housing is responsible for the support of economic development, particularly of small and medium-sized businesses and economy-related research, as well as urban development and housing. According to the legal requirements of the German federal law, municipalities and cities are responsible for urban renewal issues. The federal government is authorized to provide funding to the states for important investments of the states and municipalities, such as for urban renewal according to the Basic Law for the Federal Republic of Germany, Art. 104 b (German Federal Ministry of Justice and Consumer Protection 2014).

The Baden-Württemberg urban development promotion program supported projects in 3,125 urban rehabilitation and development areas, with a total budget of €7.3 billion between 1971 and 2018. The basis for the utilization of urban development support funding in all support programs is the Federal Building Code

TABLE 6.1 Summary of population, economy, and environment indicators for Freiburg and Germany

Indicator	Freiburg	Germany
	Population	
Population	229,144	81,800,000
Population growth per annum	0.85%	−0.08%
Population growth since 2002	+20,000 inhabitants	−693,000 inhabitants
Population density	1,497 (inhabitants/km²)	229 (inhabitants/km²)
	Economy	
Gross domestic product 2010	EU€8.752 billion	€2,643.9 billion
Gross domestic product per capita	€34,900	€32,276
Gross domestic product per labor force	€52,900	€64,344
Average economical growth	1.8% (per year)	0.9% (per year)
Debt per capita	€1,085	€25,248
Unemployment rate 2012	5.6%	7.1%
	Environment	
NOx (µg/m3)	22	42
PM10 (fine dust) (µg/m3)	18	25
Waste quantity 2011	118 kg/capita	500–600 kg/capita
Recycling rate	69%	45%
CO^2 emissions per capita	8.0t	9.0t
Drinking water consumption per capita	93 l/day	126 l/day
Ecological footprint	3.9 gha/capita	5.1 gha/capita

Source: Fraunhofer Institut für Arbeitswirtschaft und Organisation IAO, 2013

and the administrative regulations of the Ministry of Finance and Economy regarding the support of urban renewal and development measures (Ministerium für Wirtschaft Arbeit und Wohnungsbau Baden-Württemberg 2018).

Indispensable for a successful future-oriented urban development is the preparation and continuing revolution of a citizen participation-based comprehensive development concept that addresses the entire city. Essential for such a concept are the analysis of local housing stock, population development, the retail structure, basic services available near housing, education and employment provision, facilities for social services and supporting integration, as well as mobility infrastructures.

The political decision-making power is executed in Freiburg by three bodies: 1) the lord mayor, 2) the municipal council, and 3) the enacting committees. The community members directly elect both the municipal council and the lord mayor. The lord mayor is the most important person, being both the chairperson of the municipal council and head of the municipal administration. However, the municipal council is the most important institution, according to § 24 I 2 of the local government law (Landtag Baden-Württemberg 2018).

The municipal council can form consulting and decision-making committees, which are generally governed by the lord mayor or an appointed representative. The municipal council consists of 48 voluntary municipal councilors, who are elected directly by the citizens for a period of five years. The councilors are representatives of different political parties, mainly of Bündnis 90/Die Grünen, the Christian Democratic Union (CDU), and Social Democratic Party (SPD). Freiburg's municipal council is characterized by consensus across all parties regarding the sustainable and environment-friendly orientation of city policy. This was an important basic condition for the initiation and implementation of many activities in the areas of climate protection and sustainability (Fraunhofer Institut für Arbeitswirtschaft und Organisation IAO 2013).

Freiburg's land use plan needed to be updated at the end of the 1980s, particularly due the rising space demand for housing and industrial real estates. The landscape plan for the open space outside already built-up areas illustrated significant ecological risks associated with its development. Therefore, it was decided to provide the required development space inside existing neighbourhoods (Höfflin 2004).

In the 1990s, it was decided to establish two brand new districts following the city's commitment to sustainable urban development, by regeneration of a sewage irrigation field, and by conversion of a barrack area. The aim was to reduce the impact on the environment to the greatest possible degree and to create synergies and a strong balance between environmental, economic, and social sustainability aspects. Accordingly, the districts of Rieselfeld and Vauban were developed (Figure 6.2) (Fraunhofer Institut für Arbeitswirtschaft und Organisation IAO 2013).

FIGURE 6.2 Map of Freiburg showing the district areas of Rieselfeld and Vauban

Source: Adapted from Fraunhofer Institut für Arbeitswirtschaft und Organisation IAO, 2013

The Rieselfeld district development started in 1991 and was finalized in 2010. The Vauban district development started in 1993 and was finalized in 2016. The Rieselfeld district area was the property of the city of Freiburg, while the Vauban district area was purchased by the city from the Federal Government in 1993. In 1992, French troops left the barracks area after 47 years of occupation following its confiscation in 1945. Both districts are located at the border between Freiburg and the surrounding municipalities and are connected with the city's tramline network that was extended into the districts within the framework of the urban development processes (Fraunhofer Institut für Arbeitswirtschaft und Organisation IAO 2013).

Both districts areas were converted to the status of urban development measures according to the town and country planning code. Therefore these municipal developments were associated with specific requirements regarding project development and finance and were eligible to receive public funding from Baden-Württemberg state. The available funding also included a budget for participation tasks. In order to facilitate the effective and direct participation of citizens, the city established the instrument of so-called extended citizen participation. This instrument was introduced within the Rieselfeld district development framework as a result of a compensation process between nature conservation and urban development. The city and the citizen made the compromise that a bigger portion of the former sewage irrigation fields would be converted into nature preservation areas while a smaller portion would be authorized for city development, under the precondition of ongoing extended citizen participation. In order to guarantee effective participation, a citizen association with the legal form of a public association was founded. Based on the positive experiences with citizen participation integration in the district planning of Rieselfeld, a similar participation model was introduced into the Vauban district planning (Sperling 2013).

The Rieselfeld and Vauban development projects were realized with a comparatively high building density, with a floor area ratio of over one (Fraunhofer Institut für Arbeitswirtschaft und Organisation IAO 2013). Construction of single-family houses was not included in the Rieselfeld and Vauban development plans in order to avoid increased land and resource consumption. Two-family houses are the smallest building types that have been permitted in certain zones (Höfflin 2004).

The total Vauban district area is 41.3 ha and fully populated. This is approximately 1 per cent of Freiburg's 4,155.2 ha total populated area. Freiburg Vauban has a population (2016) of 5,661 inhabitants with an average age of 33.4 years that live in 2,591 households, located in 354 buildings. The majority of buildings have three and four storeys. Only 35 buildings have five storeys, and three buildings have six or more storeys. The total housing area is 185,000 m², 71.4 m² average per household, and 32.5m² per resident. The average household size is 2.2 persons, and the mean residential period is 6.5 years. The population density is 135.4 residents per hectare of populated area. This is the highest density of all districts in Freiburg with an average of 53.9 residents per hectare. The number of cars is 202 per 1,000 residents, and there is a total of 1,146 cars in the district. The car density is the lowest of all districts in Freiburg, which has 390 per 1,000 residents overall (Freiburg im Breisgau, 2017). Vauban consists of residential areas (16.4 ha), industrial areas (1.6 ha), green areas (2.6 ha), traffic areas (12.4 ha), and public spaces (2 ha) (Freiburg im Breisgau 2013).

The municipal council decided in 1993 to develop the former French barracks area in Vauban into a new urban district. In 1994 an urban development competition was put out to tender (Fraunhofer Institut für Arbeitswirtschaft und Organisation IAO 2013). In the same year, the design from the Stuttgart-based office Kohlhoff & Kohlhoff was selected from among 60 contributions and received the first prize by the jury. In the aftermath, the office further developed the competition design plan, which became the basis for the legally binding land use plan. This plan also included the streets and undeveloped areas surrounding the new district development area. In 1997, the final binding land use plan (Figure 6.3) was determined and became valid (Amt für Projektentwicklung und Stadterneuerung 2016). The construction of the first residential buildings started in 1998. The development project was completed in 2016 (Fraunhofer Institut für Arbeitswirtschaft und Organisation IAO 2013).

The starting basis for the dense, land use saving district development was the existing building block structure from Wilhelminian times. The existing structure was mixed up in a couple of rows with new buildings that had to have a minimum

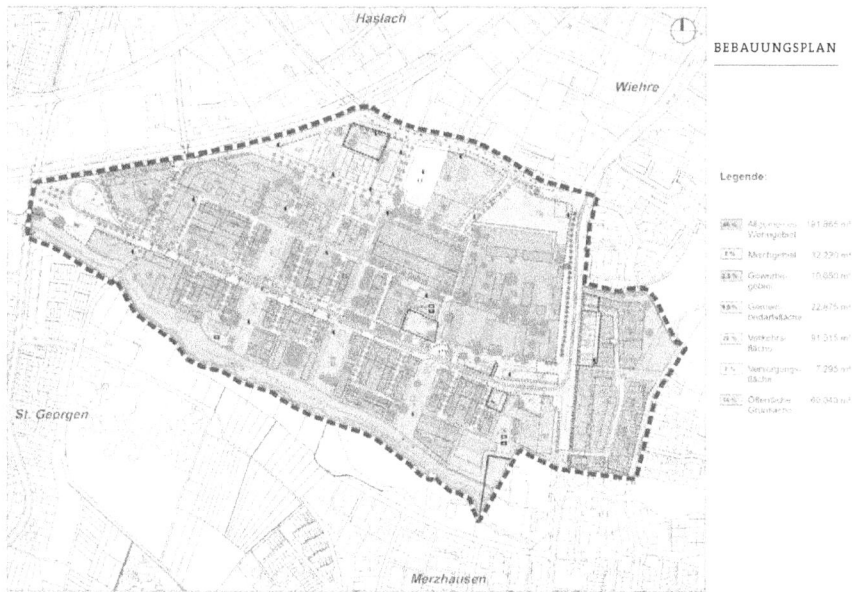

FIGURE 6.3 Legally binding land use plan for the development of Vauban

Source: Amt für Projektentwicklung und Stadterneuerung, 2016

Note: The legend denotes the different land uses, from top to bottom: general residential areas (red), mixed areas (orange), commercial areas (grey), public facility areas (purple), traffic areas (yellow), supply areas (light yellow), and public green areas (green).

distance of 19 m to the existing buildings. The maximum building height is 13 m. The whole Vauban area consists of a 20 ha residential building area, a 4.5 ha mixed-use and commercial area, a 1.7 ha community area, and a 6 ha green area (Fraunhofer Institut für Arbeitswirtschaft und Organisation IAO 2013).

Vauban was included as a key project in the federal research area "Cities of the Future". The city of Freiburg and the citizens' forum for Vauban were included as the German best practice contribution for the organization of a cooperative municipal planning process and for the exemplary engagement of citizens at the second United Nations Conference on Human Settlements, Habitat 2, in Istanbul in June 1996. The importance of the key project, Vauban, for sustainable urban development has many reasons. The development project was executed as an urban development measure. This resulted in control of the quarter's quality, the appropriate density, with a floor area ratio of approximately 1.3:1, and the effectiveness of the overall project's execution schedule. The energy consumption of the settlement was optimized by a district heating supply and the limitation of the buildings' heating energy demand to low-energy house standards (Höfflin 2004). The heating energy demand of the buildings was limited to a maximum of 65 kWh/m^{2}a (Amt für Projektentwicklung und Stadterneuerung 2016). Prematurely, it was decided to realize the biggest solar house project in Germany, with the start of construction in 1999. A socially acceptable mix of alternative, student, and family-oriented forms of housing, as well as a mix of rental and owned housing, such as multi-storey residential buildings, student apartments, row and town houses, was provided. The housing in the entire district is in calmed traffic areas and without car parking places. Car parking is centrally provided in a multi-storey parking garage with an integrated photovoltaic power plant. In 2006 the district was connected to Freiburg's tramline network. The distances between the different uses, housing, shopping, service facilities, and workplaces, are short (Höfflin 2004).

The urban sustainability challenge: Creating ecological mixed-use districts

A big part of Freiburg was destroyed during the Second World War. After 1945, the city centre of Freiburg was reconstructed according to the original urban plan and building typologies. In contrast to the paradigm of the car-friendly city, which would have required the construction of adapted road infrastructures, a tram network was established as the backbone of the city's public transport system. Accordingly, a car-free inner city, functioning as the centre for the whole city could be established, and the original small-scale and versatile urban structure could be preserved. Due to a high influx of refugees and population growth, Freiburg was confronted with a significant housing shortage. In order to cope with this shortage, occupation of apartments was regulated until the mid-1960s. The maximum floor area for a four person household was limited to 45 m^2. Apartments that exceeded the maximum area per person needed to be occupied by additional persons. In 1964, the municipal council decided on a five year social housing program, which resulted in the fast construction of big car-friendly apartment settlements in the periphery of Freiburg. In contrast to the car-oriented developments in the city's periphery, the city centre was completely closed for car traffic from 1973. This was also the year when the first bicycle path concept for the city was developed (Fraunhofer Institut für Arbeitswirtschaft und Organisation IAO 2013).

In the 1980s, the continuing housing shortage resulted in citizen protests. Car-friendly designed suburban areas continued to grow. In order to reactivate the existing city district centres and to keep essential retail services, a decentralized so-called centre and market concept was developed. Furthermore, politics and urban planning resulted in the building of new city districts in order to return economic power from the surrounding municipalities back to Freiburg. The development objectives of the new districts were to address and answer social, energy, and environmental questions. Only three from 33 planning areas were designated for development because real estate issues prevented development in the majority of areas. The Chernobyl nuclear reactor catastrophe in 1986 resulted in a mainstream anti-nuclear energy movement. Freiburg's politics reacted with the first energy supply concept for a German metropolis, and the introduction of a low-energy house standard, a compulsory municipal energy performance certificate (EPC) with a limitation on the heating energy demand for new building construction (Fraunhofer Institut für Arbeitswirtschaft und Organisation IAO 2013).

The mass influx of population in the 1980s required the provision of affordable housing and of workplaces. Freiburg decided to develop a new district in the form of an urban quarter with distinctive forms of multi-storey dwellings, sustainable water management, energy savings, a mix of different social classes, and a primary connection to the city's tram network. In 1991, it was decided to develop this district on the former sewage farm area of Rieselfeld (Fraunhofer Institut für Arbeitswirtschaft und Organisation IAO 2013). The development of the Rieselfeld district was realized as a low-energy settlement. All buildings in this new urban district had to be connected to the city's district heating network. Furthermore, the heating energy demands of the buildings were limited to a maximum 65 kWh/m^2a. The residential construction costs were partly supported by the state of Baden Württemberg (Institut Wohnen und Umwelt 1996).

In 1992, France withdrew their troops from the 40 ha barracks area of Vauban, which is located within the city of Freiburg. In 1993, the city purchased 34 ha of Vauban in order to develop a sustainable district. Based on the sustainability-oriented principles that were formulated in the 1980s, concrete development goals for the district were formulated. The main goals were the integration of business, residential, and mixed-use areas, in a traffic-calmed district with a comparatively small number of cars. Furthermore, small parcelled properties were to be primarily marketed to individual building owners and building owner communities. The integration of decentralized rainwater management measures as well as the passive and active use of solar energy was high on the agenda (Fraunhofer Institut für Arbeitswirtschaft und Organisation IAO 2013).

The planning innovation

Innovation process

From the beginning of the 1990s, Freiburg's municipal government adapted the development tools needed for a comprehensive sustainable urban planning strategy. The strategic concept is based on three principles: 1) a compact city, 2) a mix of different social classes and participation, and 3) a climate-responsive city. Responsibility for the transfer of the overall concept into concrete city development lies in three conceptual domains: 1) market and centre, 2) building area-related policy principles, and 3) climate responsive city principles.

The principal aim of the market and centre concept is the development of a vibrant compact city with short distances and the reversal of the relocation trend of retail activity from city centres to car-friendly locations in the periphery. The main instruments of the successful concept that was introduced in 1992 are the centre hierarchy and a sales mix list. According to the differentiation of fully, partly, and non-integrated locations for retail, four area types were designated, which are subject to different regulations. Integrated locations for retail activity are the central city centre and urban district centres for local supply. The sales mix list differentiates the complete retail sales mix in assortments that are centre relevant and not center relevant. The market and centre concept is applied by determinations in legally binding land use plans.

The building area policy principles were adopted by the municipal council in 1998 and revised in 2012. These principles are the central instrument used to fix social, ecological, and financial sustainability targets in urban development. The building area-related policy principles secure the consideration of similar principles for financing as well as for the implementation of social and ecological standards in all general town planning schemes, and for the composition of sale contracts for municipal building plots. The most important aspects of the innovative process are (Fraunhofer Institut für Arbeitswirtschaft und Organisation IAO 2013):

- All municipal services related to building law need to be refinanced by the beneficiaries. This includes, for example, planning services, expert opinions, infrastructure connection facilities, and the provision and management of compensation areas.
- The construction costs for required kindergarten places in a planning area need to be defrayed by investors.
- Investors are pledged to realize 30 per cent of newly built housing areas in the form of subsidized rental or owned housing constructions. Alternatively, the city will receive 10 per cent of the areas for subsidized housing construction measures.
- All developments need to include a minimum portion of housing and mix of different apartment sizes.
- All new building constructions have to comply with the regularly updated and strengthened municipal energy efficiency standards. By 2013, the municipal standard limited the maximum primary energy demand of residential buildings to

55 per cent of the national Energy Saving Ordinance EnEV2009 (Bundesinstitut für Bau- Stadt- und Raumforschung im Bundesamt für Bauwesen und Raumordnung – BBSR 2015).

- Building areas of special significance for urban developments need to be reserved for specific required functions, such as local supply of goods, services, or social infrastructure.

The contents of the building area-related policy principles are transferred, for private areas, to urban land use planning, by binding urban development contracts between city and investor. For public areas, the principles are included in the sales contract. The building area-related policy principles facilitate a strategic long-term perspective of the real estate policy for the achievement of sustainable development targets (Fraunhofer Institut für Arbeitswirtschaft und Organisation IAO 2013).

The participation of Freiburg's population in planning practice is a major innovative component of the land use planning. In addition to public practices that are regulated by the town and country planning code, the participation of all concerned persons is realized in different ways, such as public debates and planning workshops. Furthermore, engaged citizens play very important roles in the development of urban districts. The participation of the citizenship is organized and moderated by volunteers who live in the concerned districts. The city supports this process with information, materials, and methodological advice. The results from the participation processes are implemented by the municipal urban development council in a planning document (urban district guidelines) that is discussed and agreed with the city council and the citizenship in the concerned districts. The municipal council adopts the urban planning guidelines as standards for future planning (Fraunhofer Institut für Arbeitswirtschaft und Organisation IAO 2013).

Successful project implementation requires solid conceptions and is influenced by external impact factors. Some factors and their impacts on the project development are known from the beginning of a project. Their continuous influence on a project can be already considered during the project planning. Other impact factors occur later in the project implementation process and eventually require an adaption of the project. Both impact factor types can function as drivers or barriers to a project. Forty-three impact factors with varying influence on successful project implementation were identified in Freiburg's sustainable urban development projects. The factors were assigned to the following 12 categories in order from high to low occurrence (Fraunhofer Institut für Arbeitswirtschaft und Organisation IAO 2013):

- Competency/Experience/Motivation
- Cooperativeness of involved project stakeholders
- City-specific basic conditions
- Acceptance of citizens
- Strategy/Communication/Image
- Demands/Requirements (e.g. social benefits, liveability, security)

- Urban resources (financial, human) and capacity to act
- Investors/Financing/Business models/Incentives
- Organization/Administration structures/Projects
- Decision-makers/The public
- Regulations
- Technologies/Products

Drivers in Freiburg that have a high impact on other factors and are only to a very limited degree influenced by other impact factors, are (Fraunhofer Institut für Arbeitswirtschaft und Organisation IAO 2013):

- Leadership personalities who promote projects and processes, and motivate other stakeholders. These are primarily the lord mayors of the last decades, Dr Böhme and Dr Salomon, but also game-changing department directors or private innovators. The architect Rolf Disch, for example, stimulated sustainable architecture development and became internationally famous with plus energy buildings and the solar district in Vauban.
- Municipal council committees give detailed attention to specific topics. Four of 14 relevant municipal council committees have a clear reference to sustainable urban development. These are the committees for construction, urban development, environment, and transport.
- Municipality's shareholdings in service providers, and municipal corporations, such as the municipal energy provider Badenova. This facilitates the city to develop and implement its own strategies via its service providers.
- The presence of scientific competency and institutions is a strong driver of sustainable development because the city can use local expertise for the development of sustainability strategies and innovations. Furthermore, many experts are engaged as competent citizens and campaign for ambitious solutions.
- The composition of Freiburg's society works also as a continuous driver for the sustainability process. The society consists of many employees in education, science, and research. The city has a strong service provider sector and only limited industry.

Levers in Freiburg that have a strong impact on other factors but are also passively influenced by many other impact factors, are (Fraunhofer Institut für Arbeitswirtschaft und Organisation IAO 2013):

- The municipal council, which has high significance because it is the most important urban decision-making committee.
- Continuity and a long-term perspective on activities that promote sustainability are the second lever of Freiburg's sustainable urban development. The first energy concept and the first low energy standard, introduced in 1986 and 1992, have been continuously further developed. In addition, since the beginning of the 1990s the building land-related policy principles have been

respected as objectives for the development of new areas. The market and centre concept has shaped the development of urban district centres and the city centre over the same period. The long-term continuous pursuit of sustainability objectives with coherent strategies is one of the most important success factors of sustainable urban development.

The following enablers in Freiburg are influenced by many other impact factors, have only little influence on other factors, and are crucial for the success of projects (Fraunhofer Institut für Arbeitswirtschaft und Organisation IAO 2013):

- The mobility behaviour of citizens, which is dependent on available mobility modes and many more influential factors, is essential for the success of mobility projects, such as the use of public transport and car sharing.
- Sufficient funding is an important success factor for sustainability projects. The reservation of income for the financing of sustainability and climate protection measures, such as the energy efficient renovation of the building stock, is realized by appropriation of financial resources reserved for the renovation of the building stock and from license fees of Badenova.
- The availability of housing has a direct influence on the development of rents, social integration, and market prospects of new housing. This factor has the highest disparity because it has a big influence on urban development, but it is difficult to direct due to the very limited availability of new development areas.
- The availability of venture capital investors is crucial for the initiation and the success of innovative sustainability projects. Badenova's investment in sustainable climate protection and water management projects via its innovation fund lowers barriers for additional venture capital investments.
- The municipal growth strategy in future-oriented sectors facilitates sustainable development by the local promotion of economic development, particularly in the environment and solar economy sectors.
- Self-imposed principles in policy and administration define ongoing strengthened minimum standards for sustainability-related actions that become the basis for new developments. This results in the strong sustainable development of Freiburg. Examples are being a signatory to the Aalborg commitments, the adoption of strengthened climate protection targets, the building land policy principles, and the low energy building standard.

In addition to these outstanding impact factors, other factors with negative impact on other impact factors can be identified. These factors can only be influenced to a limited degree. The factors in Freiburg are the availability of affordable housing, population growth, and the dependence on legal requirements. A big number of regulations, e.g. in the areas of urban and spatial planning and building design, at state or federal levels, can complicate or counteract municipal policy. Excessive population growth might nullify the progress made, e.g. in housing construction, land consumption, and improvements in efficiency. Accordingly, the legal objectives might not be achieved. Moreover, demographic change is both a driver and a system risk (Fraunhofer Institut für Arbeitswirtschaft und Organisation IAO 2013).

Trade-offs

This section sets out the processes that have prevented negative trade-offs between ecological, environmental, and social sustainability aspects in the urban developments within Freiburg. This city is famous for being Germany's ecological capital due to different planning and development projects that have been realized by the municipality in close cooperation with a committed private sector. Due to an effective public-private partnership, the city became a pioneer in ecologically oriented urban planning. Landmark activities resulted in participation in the federal research project "City of the Future" and in Freiburg becoming a reference city for sustainable urban development. Some of the most important activities are (Höfflin 2004):

- The integrated and citizen participation-based urban development of the conversion area of Vauban.
- Dialogical process-based integrated planning for future land use.
- Planning and implementation of holistic mobility concepts.
- Leadership in passive and active solar energy technology development and application.
- Highly energy efficient building developments, including Passive Houses.
- Development of combined living and working environments at neighbourhood and district levels.
- Entrepreneurship of individual citizens and of building owner communities.

The project processes for sustainable district and city development in Freiburg can be divided into five main project phases. These are initiation, decision, planning, implementation, and evaluation. The most important findings from the analysis of the sustainable urban development projects in Freiburg, including Rieselfeld and Vauban, are summarized below.

Initiation phase: drivers for the initiation of projects can be found in all actor groups, particularly in an engaged and ambitious population and in active interest groups. Their engagement is also reflected in the community council and represented by a city administration that is continuously and with a long-term perspective working on sustainability topics. The urban development projects of Rieselfeld and Vauban have been initiated due to the demand for housing. The ecological orientation of the development projects was based on the city's energy policy objectives that were determined previously and based on the anti-nuclear movement in the aftermath of the Chernobyl accident. Numerous interest groups, engaged actors, such as planners and architects, and users, primarily ensured the ecological consistency achieved in project implementation.

Decision phase: Rieselfeld and Vauban are municipal development projects. The municipal council is, therefore, the pertinent decision-making committee. The lord mayor and the administration developed the projects and prepared relevant presentations, and have, accordingly, a significant influence on the decisions of the municipal council. However, the municipal council made decisions in sustainable development areas that went beyond the administrative presentations. Municipal council committees prepare the content of decisions. Therefore, the committees consist of members with generally good sustainability-related expertise. The continuous steering of sustainable urban development projects by municipal council committees proved to be very valuable. The municipal council committees' decisions regarding the structure and funding of projects have been groundbreaking. The lord mayor proposed a new financing concept for the municipal development projects, which was approved by the municipal council committees. The financing concept was based on the introduction of self-financing schemes for development measures and independent project groups.

Planning phase: the planning of the city districts was based on extraordinarily comprehensive participation processes that actively involved all relevant actor groups. The foundation and public financing of the registered Vauban Urban District Association supported the organization of potential future citizens' interests. Urban development competitions were executed for both sustainable urban development projects. Particularly relevant in this phase were the specific competition objectives and the decision to realize a flexible development concept with a diversity of investors, building types, building uses, and the ongoing adjustment to a participation-based learning planning.

Implementation phase: outstanding project management, application of the learning planning method, and continuous exchange with the municipal council proved to be particularly valuable during the implementation phases. Autonomous project teams, independent budgets, and direct access to both the lord mayor and the building mayor facilitated the action capabilities of the project management. The learning planning principle facilitated flexible feedback for changing basic conditions during the implementation phase.

Evaluation phase: the results of Freiburg's sustainability policies are measured by the preparation of a biannual climate protection balance. The results are discussed in the municipal council. In 2013, a so-called sustainability compass, an innovative steering tool for the evaluation of sustainable development impacts, was introduced in Freiburg. The high demand for apartments with rising prices and vibrant city life prove the success of the urban development measures in Rieselfeld and Vauban (Fraunhofer Institut für Arbeitswirtschaft und Organisation IAO 2013).

Conclusion

Based on the experience in Freiburg, the following conclusions can be drawn for the potential transfer and replication of the innovations to other municipalities and city contexts.

The setting of clear, regularly reviewed, and adapted objectives is a premise for successful sustainability policy. Sustainability policy needs to have a long-term perspective. Sufficient economic funding, in Freiburg, for instance, by a concession levy appropriation of 10 per cent, is important for sustainability policy development. Sustainability needs to be appropriately positioned in the municipal administration. The general public and municipal enterprises need to be permanently engaged and motivated. Active awareness raising and image building are preconditions for successful sustainable development. Cooperation with the region is indispensable, particularly in the area of mobility.

Continuity of industry, municipality, and civil society cooperation is important in order to facilitate the development of new and innovative technological solutions. Innovative solutions and pilot projects can be developed with the cooperation of research institutions and innovators. Customer-oriented integrated solutions that are cooperatively offered by multiple industries and trades are helpful to overcome customers' commissioning barriers. In combination with attractive financing schemes, customers' uncertainty, such as regarding the selection and commissioning of the right solution for the energy efficient renovation of buildings, can be overcome. The support of regional innovative sustainability projects can support the innovation potential of a region, improve its image, and facilitate the identification and assessment of new technologies. Active networking of private industries with public service providers facilitates the implementation of integrated multi-modal mobility concepts, including public transport, bicycle, and car sharing (Fraunhofer Institut für Arbeitswirtschaft und Organisation IAO 2013).

The interregional recognition of the Vauban project results particularly from the holistic urban development project concept. Projects of comparable size are generally more realizable in expanding large cities than in small shrinking cites. Furthermore, the Vauban development process was significantly influenced by specific local basic conditions. Accordingly, the potential transferability of the Vauban project to other urban development projects generally concerns single aspects of the policy principles, project modules, and planning processes (Höfflin 2004) that have been discussed within this chapter.

References

Amt für Projektentwicklung und Stadterneuerung (2016) *Quartier Vauban – Von der Kaserne zum Stadtteil*, City of Freiburg im Breisgau, Freiburg

Bundesinstitut für Bau- Stadt- und Raumforschung im Bundesamt für Bauwesen und Raumordnung – BBSR 2015 Energy Saving Ordinance (EnEV) (2015) (www.bbsr-energieeinsparung.de/EnEVPortal/EN/EnEV/enev_node.html) accessed 9 March 2018

Fraunhofer Institut für Arbeitswirtschaft und Organisation IAO (2013) *City Report Freiburg*, Fraunhofer Institut für Arbeitswirtschaft und Organisation IAO, Freiburg

Fraunhofer Institute for Solar Energy Systems ISE (2018) *Fraunhofer Institute for Solar Energy Systems ISE* (https://www.ise.fraunhofer.de/en.html) accessed 9 March 2018

Freiburg im Breisgau (2013) *Statistisches Jahrbuch 2013*, Amt für Bürgerservice und Informationsverarbeitung der Stadt Freiburg im Breisgau, Freiburg

Freiburg im Breisgau (2017) *Statistisches Jahrbuch 2017*, Amt für Bürgerservice und Informationsverarbeitung der Stadt Freiburg im Breisgau, Freiburg

Freiburg Wirtschaft Touristik und Messe GmbH & Co. KG (2016) *Green City Freiburg*, Freiburg Wirtschaft Touristik und Messe GmbH & Co. KG, Freiburg

Freiburger Agenda 21 (2018) *Agenda 21-Büro Freiburg* (www.agenda21-freiburg.de) accessed 9 March 2018

German Federal Ministry of Justice and Consumer Protection (2014) *Basic Law for the Federal Republic of Germany in the revised version published in the Federal Law Gazette Part III, classification number 100–1, as last amended by Article 1 of the Act of 23 December 2014 (Federal Law Gazette I p. 2438)*, German Federal Ministry of Justice and Consumer Protection, Bonn

Höfflin, P. (2004) *Werkstattbericht "Nachhaltigkeitsindikatoren" für die Stadt Freiburg*, Amt für Statistik und Einwohnerwesen der Stadt Freiburg im Breisgau, Freiburg

ICLEI – Local Governments for Sustainability – European Secretariat (2018) *The Aalborg Commitments* (www.sustainablecities.eu/the-aalborg-commitments/) accessed 9 March 2018

ICLEI European Secretariat (1998) *ICLEI Europe* (http://iclei-europe.org/home/) accessed 9 March 2018

Institut Wohnen und Umwelt (1996) *Eine Geschichte der Niedrigenergiehäuser bis zum Passivhaus*, Institut Wohnen und Umwelt, Darmstadt

Landtag Baden-Württemberg (2018) *Gemeindeordnung Baden-Württemberg. GemO, vom 24.07.2000* (www.landesrecht-bw.de/jportal/?quelle=jlink&docid=jlr-GemOBWpG1&psml=bsbawueprod.psml&max=true) accessed 9 March 2018

Ministerium für Wirtschaft Arbeit und Wohnungsbau Baden-Württemberg (2018) *Ministerium für Wirtschaft Arbeit und Wohnungsbau Baden-Württemberg* (https://wm.baden-wuerttemberg.de/) accessed 4 March 2018

Ökostation Freiburg (2018) *Ökostation* (www.oekostation.de/com/) accessed 4 March 2018

Sperling, C. (2013) Keine Wirkung ohne Risiko – Ein Bericht aus Vauban/Freiburg, *Planung neu denken*, 2(3), 10.

7

SUKUNAN VILLAGE, YOGYAKARTA, INDONESIA

Environmental sustainability through community-based waste management and eco-tourism

Iswanto, Sita Rahmani and Sonia Roitman

Introduction

Increasing waste, one of the negative impacts of human activities, urgently needs to be addressed in the pursuit of sustainable development. The United Nations Agenda 2030 that sets the basis for the Sustainable Development Goals (SDGs) identifies reduction and recycling of waste as a target: 'By 2030, substantially reduce waste generation through prevention, reduction, recycling and reuse' (Target 12.5, UN 2015). Waste management requires coordinated efforts at the local level and can be done by the public sector (local government), private sector and civil society (organised communities and non-governmental organisations (NGOs)). This chapter analyses a successful innovative initiative of solid waste management in Sukunan, a village located in the peri-urban area of Yogyakarta, Indonesia. This is a project designed and driven by the local community in a collective and organised form since 2004. It has not only been sustained and improved for 13 years, but also replicated by other communities in Yogyakarta Province and Indonesia.

Sukunan village, Yogyakarta

Sukunan is a village in Gamping sub-district, Sleman Regency, Yogyakarta Special Province, Indonesia. It is located in the Metropolitan Area of Yogyakarta (which has a population of 1.5 million, while 4 million people live in the province, BPS 2017). Sukunan is situated in a peri-urban area, approximately 5 kilometres west from Yogyakarta city centre, combining residential land use with farming and agriculture (see Figure 7.1).

Based on Indonesian administrative division, Sukunan is the lowest administrative governmental unit (village). It is organised in five neighbourhoods. In 2016, 253 families (1,117 people) lived in the village (Sukunan Village 2017). It occupies

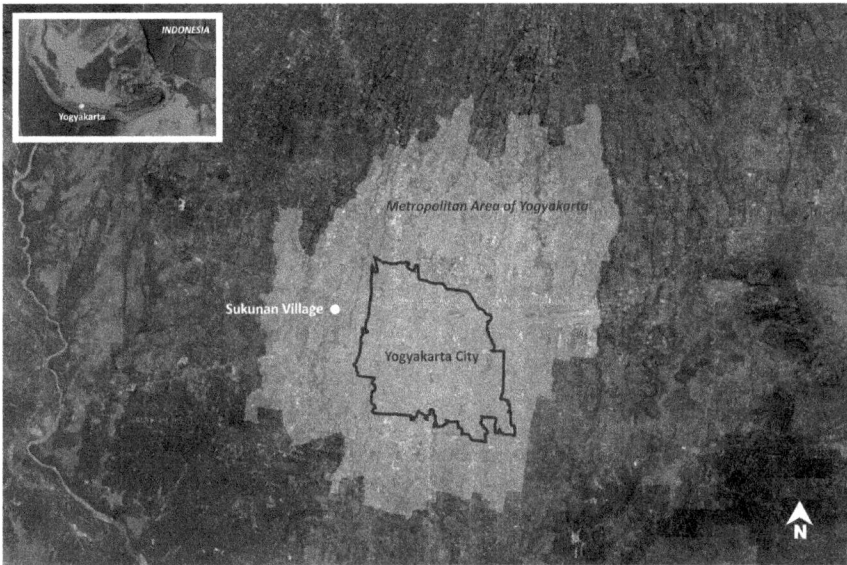

FIGURE 7.1 Location of Sukunan

Map by Sita Rahamani based on Google Map 2018

42 hectares (Razak 2010). Agricultural activities (especially rice fields) are the main livelihood for residents. The majority of household heads have only finished primary school and have low-skilled jobs (farmers, construction workers, street vendors, small-scale entrepreneurs and low-skilled public or private sector employees) with incomes below the minimum wage in Yogyakarta.[1]

The planning system is decentralised in Indonesia, with the local governments being responsible for giving planning permits and regulating land use, according to local and provincial urban plans (Law No. 26/2007 about Spatial Planning, Government of Indonesia, 2007). The local government elaborates local development plans that include sectoral targets, such as infrastructure, health and education facilities for specific areas, following national and provincial plans. Municipalities are organised in local departments. In Sleman, the Department of Public Works is responsible for housing, roads, bridges and the provision of clean water. Sanitation is managed by the Department of Health, while waste management is the responsibility of the Department of Environment.

In Sleman Regency, not all areas are well serviced due to the lack of financial resources. In the case of Sukunan, there were about 40 households who did not have toilets and defecated in open areas such as small rivers and irrigation canals in 2004. Dian Desa Foundation, a local NGO, conducted an examination and confirmed high concentration of E. Coli bacteria around wells near the river and irrigation canals (Tanaka 2008). Most families in Sukunan use underground water from their own wells that is boiled before drinking. The majority of roads are paved while alleys are dirt roads. Before this community project started there was no organized waste disposal system in Sukunan.

The urban sustainability challenge: Solid waste management

Solid waste management constitutes a challenge for urban planning and sustainability concerning not only environmental aspects but also economic and equity aspects. In regards to environmental issues, our high-consumption society produces 1.3 billion tonnes of waste per year[2] and this is expected to increase to approximately 2.2 billion tonnes per year by 2025 (Hoornweg and Bhada-Tata 2012). Increasing waste is difficult to manage and pollutes oceans, rivers and air. Household solid waste has become a serious concern in urban areas (Dhokhikah et al. 2015).

In relation to equity, waste disposal affects most the residents of the cities of the Global South where there is limited capacity for formal solid waste management systems to deal with the waste produced (limited disposal locations and treatment plants and limited financial resources) and the system heavily relies on informal practices (Wilson et al. 2006). Additionally, poor groups are more vulnerable to the effects of weak waste management because they are not able to afford the payment of formal solid waste management services and therefore are not able to properly manage the waste in their neighbourhoods (WHO 2009). Most slums and informal settlements have no adequate system of garbage collection and disposal (UN-Habitat 2016). Although informal solid waste management practices, such as those by scavengers, constitute an important source of income, they expose people to high health risks (Colon and Fawcett 2006).

The United Nations Agenda 2030, which is the mainstream guideline document on development, recognises the importance of treating waste to create healthy and clean human settlements. Within the SDGs, Target 12.5 refers to 'substantially reduce waste generation through prevention, reduction, recycling and reuse' by 2030, contributing to achieve SDG 12 ('Ensure sustainable consumption and production patterns') and SDG 11 ('Make cities and human settlements inclusive, safe, resilient and sustainable') (UN 2015). Target 12.5 then relates to the three planning priorities of equity and social justice, environmental protection and economic development (Campbell 2016). Given that the SDGs have been agreed by all United Nations country members, planning programs, projects and activities should aim to contribute to their achievement.

Global waste management has recently become a highly discussed topic, as there is an increasing awareness of the enormous challenge it represents. Environmentally, planet Earth has no capacity to manage the produced waste, especially plastic. Levels of contamination and irreversible environmental damage continue to increase. Politically, countries that were receiving overseas waste such as China are no longer willing to do this. Hence, the only solution is for waste to be managed at the local scale, becoming mainly the responsibility of local governments and communities. There have been formal and informal practices implemented at the local level. In the Global South, there is a mix of both practices (Wilson et al. 2006).

The practices of the Zabaleen (garbage collector communities) in Cairo, Egypt, are well known. They started in the 1980s, although there had been waste collecting practices in the area as a livelihood strategy for nearly a century, and became a practice to be replicated in other countries. The Zabaleeen are rural migrants who maintain strong kinship and community bonds. They live in settlements on the urban fringe of Cairo. They collect and sort waste and then sell inorganic waste to intermediary buyers who later sell the waste to large companies. Organic waste is used to feed animals, which are also an important source of income for the Zabaleen. They informally handle one third of Cairo's garbage and have improved the capability of the city to manage waste, at no cost to the local government (Fahmi and Sutton 2006). In India, in the early 2000s, a local NGO initiated the 'Zero waste management system' program promoting community-based involvement in some upper- and middle-class neighbourhoods of Chennai and Hyderabad. It consisted of door-to-door household waste collection and the sorting of material (80 per cent organic and 15 per cent recyclable). Thus only the remaining 5 per cent of the household waste required handling by the local government. Households paid a fee to a community organisation for the collection service that was used to hire workers who collected and sorted the waste. The local government provided land and infrastructure for composting (Colon and Fawcett 2006). In Addis Ababa, Ethiopia, in the early 2000s, the local government created microenterprises to run door-to-door waste collection activities with the double aim of improving the municipal waste collection system and providing job opportunities as a vehicle to reduce the high level of unemployment. The system has received some criticisms, such as the failure to integrate formal and informal waste collection practices (Baye Alene 2018).

Indonesian cities produce about 200,000 tonnes of waste every day and nearly half of this is produced by households (Wijayanti and Suryani 2015). In Yogyakarta Province, the average production of waste per day is 644 tonnes. Only 65 per cent of that can be handled by the government (Public Works of D.I. Yogyakarta Province 2016).

In Indonesia, most waste is dumped either in officially designated areas or in illegal dumps, including roadsides and watercourses. Common issues of solid waste management are low commitment of local government to prioritise solid waste issues, scarce funding availability, and lack of proper institutional arrangements, public awareness and law enforcement (Mursito et al. 2013). Most of the waste is organic and can be used for compost while the non-organic components of the waste have market value for reuse or recycling (MacRae and Rodic 2015). Thus, there are many people who work as scavengers (sorting out waste in disposal sites, searching for items that can be sold or recycled). These informal livelihood practices imply a reduction of between 9 and 15 per cent of waste and benefit the city, saving money on collection, transport and disposal of waste and also cleaning areas and extending the lifetime of disposal areas (MacRae and Rodic 2015). The approach to manage solid waste management has changed in Indonesia since 2008 from an 'end-of-pipe' approach (management of the dumped waste) to 'reduction of the source' (waste produced by the household). Cities and residents are encouraged to optimise waste reduction in every stage of waste processing, not only at the final disposal site (Mursito et al. 2013). One of the main challenges in relation to solid waste management is how to recycle or reuse most of the waste, without just 'cherry-picking' recycling material and leaving unmanaged large portions of waste. Thus, the learning of practices for handling and managing the waste becomes a key for the success of any solid waste management system.

Yogyakarta City, Sleman Regency and Bantul Regency are the three municipalities that form the Metropolitan Area of Yogyakarta. They have joined efforts to manage solid waste. Each municipality provides funding proportional to the waste disposed in the final disposal site. While this cooperation has been running since 1997, several issues have emerged to challenge improvement, such as institutional agreements between local governments, determining roles and responsibilities, financial contributions, continuous access to budgetary funds for operation and maintenance, and obtaining land that can be used as disposal sites (Friedman 2013). Economic and financial agreements become complicated. Collecting and processing waste is also an expensive service for local governments (Bohm et al. 2010). Indonesia spends between 80 and 90 per cent of the municipal solid waste management budget in collection costs (Hoornweg and Bhada-Tata 2012). In the case of Yogyakarta City for example, the local government paid IDR$1.4 billion (US$104,000) in 2015 for the solid waste to be taken to the final disposal site (Rusqiyati 2015).

Until the late 1990s, there was no solid waste management system in Sukunan. Waste was burnt or disposed anywhere in the village (irrigation canals, vacant land, and kerbside) and created a dirty, smelly and unhealthy environment. This also created conflicts with nearby farmers who complained about the increasing amount of garbage dumped in their fields. Plastic waste disrupted rice growth and damaged the rice. Sukunan villagers did not know how to manage waste and were not aware of the health risks associated with unmanaged waste.

The planning innovation: Community-based waste management

In 1997, the first author of this chapter (hereafter referred as 'project originator') moved from Yogyakarta City to Sukunan. After experiencing the problems associated with the lack of a solid waste management system in Sukunan, he decided to start implementing some practices for waste management. In 2002, using his knowledge and expertise as an environmental health practitioner, he developed a series of practical innovations to treat waste. First, he built a clay barrel composter to be used at home, allowing families to make compost and use it as fertiliser or sell it to local farmers. Second, he started separating waste at home, based on what he had learned from the practices developed by scavengers who were trading sorted waste to wholesalers. Later his wife started to develop handcrafts (wallets and shopping bags) re-using and recycling waste that could not be sold to factories, such as food packaging.

The project originator's family developed and tried these practices at home for about four months in 2002–2003. Once he was convinced of the feasibility of these practices, he started a process of dialogue with community leaders and other community members, creating awareness of the need to implement a waste management system in Sukunan. Initially only a small number of residents, including one of the community leaders, was interested in the project. Some villagers did not believe the project could be successfully implemented. There were concerns about the financial sustainability of the project and the difficulties of changing community practices. However, the main concern was about the project approach. The community was reluctant to participate in a top-down project that considered the community as 'an object of development'. There were several community meetings for residents to discuss ideas about the project and how they would like to implement it. It was essential for the community to understand that the project was driven by the community and the community itself would be making decisions about project development. This dissolved the initial resistance to the project. The residents were also attracted to the potential economic benefits from their waste management activities.

The project required financial support to commence. The community leaders tried unsuccessfully to get support from the local government and the private sector. Finally, at the end of 2003, a private donor from Melbourne, Australia, provided the required funding from the project to be started (only US$500). Although this funding is not a large amount for such a project, this private donation was essential to get the project started.

Innovation process

Between January and April 2004, the project originator and other residents worked on five main activities including: 1) community consultation through focus group discussions on the priorities and needs related to solid waste management practices; 2) establishment of a waste management team, which designed a plan for action in consultation with the community (this means that the plan was not done just by the team, but in consultation and collaboration with the community); 3) establishment of local rules for solid waste management; 4) community awareness and training on waste management practices; and 5) provision of solid waste management facilities (the community youth made 180 rubbish separation hangers and 66 rubbish bins). Thus, the project is based on the 3Rs (Reduce, Reuse and Recycle), considering that change needs to start at the household level and be later expanded to the community and city levels.

The project was launched on 25 April 2004. Sukunan community, representatives from local universities, the local media and donors attended the event. Since then the core of the project has remained the same, but over time some activities have been added. Currently the project consists of four main activities: 1) separation and disposal of solid waste; 2) reuse and recycling of waste for handcraft-making; 3) production of bricks with Styrofoam waste; and 4) training and raising awareness on sustainable solid waste management practices, including visits to the village by outsiders.

The separation of solid waste, the first activity, is done by each household in the community. Families that were initially not interested in the project slowly became involved. There were three factors that were identified as a cause of change in the community. First, 'predisposing factors' that showed that the increasing knowledge and awareness after several community meetings on the negative effects on health of untreated waste had turned around residents' perceptions. The community also noted the concrete results of the project through the use of recycled material. Second, 'enabling factors' related to the availability of facilities for sorting garbage and waste processing. The installation of sorting bins and learning on how to prepare compost made it easier for residents to start implementing these practices. Third, 'reinforcing factors' that were triggered with the elaboration of local guidelines for the community to manage the project and the overall waste management process, the support received from the community leaders and the economic benefits experienced by households once they started waste management practices.

Currently 85 per cent of the households in the village separate their waste. Four types of rubbish are separated: plastic, paper, glass and metal, and organic kitchen waste. Each household disposes the garbage in the recycling containers located around the village (there are currently 63 bins in the village, located in 21 'spots' – see Figure 7.2). Bins are emptied weekly and a collector takes the waste to Sukunan's garbage depot. Rubbish is sold to the recyclable goods traders bimonthly. The average waste sales are IDR$275,000 per month (about US$20), with IDR$100,000 used to pay part of the salary of the garbage collector[3] and the remaining IDR$175,000 deposited in the village fund. Each household also makes compost from the organic kitchen waste that is collected in clay pots. Garden waste and cattle manure is also used for compost made collectively by a group of residents. About 1.5 tonnes of compost is produced per month in the village and 60 per cent is used as fertiliser by the community and 40 per cent is sold to buyers outside Sukunan (the approximate income from this activity is IDR$480,000 per month – US$$35). The untreated waste is about 15 per cent of the total and is sent to a landfill as residual waste. The hazardous and toxic waste is collected and stored temporarily in the village and later handed over to the local government of Sleman, responsible for managing this type of waste.

The second activity is the manufacturing of handcrafts. A group of women started a project to make handcrafts with recycled food and beverage packaging collected in Sukunan. The group also buys recycled material from school canteens, small shops and restaurants outside the village. The group produces small items such as wallets, handbags, hats and school folders, which are sold to tourists visiting Sukunan. The income received from the sale of each item goes to the person who made the item (70 per cent) and some is allocated for buying materials (25 per cent). The remaining 5 per cent of this sale goes to the village fund. There are currently 19 women involved in handcraft making. Each one gets between IDR$750,000 and IDR$1,250,000 (US$55–92) per month from this activity. Progressively other recycled and reused items such as fabric, egg shells and old car

FIGURE 7.2 Sorting bins in Sukunan

Photo by Sonia Roitman

tyres have been added to plastic and packaging for the making of new items such as cushions, plates, aprons and re-usable cloth sanitary towels for women.

In May 2006 there was an earthquake in Yogyakarta and some houses in Sukunan were damaged. A group of seven young local residents, who had received some training on waste management practices in the village, started to fabricate bricks with Styrofoam garbage, which became the third activity of the Sukunan project. The bricks are made from one part cement, three parts sand and three parts granulated Styrofoam. The material mixture is inserted into an aggregate, pressed and dried in the sun. The bricks are used to build walls and were used to rebuild five houses of earthquake victims and later to build public facilities. The community also makes pots with Styrofoam, which are for plants to decorate the main streets of the village.

The fourth activity of the project consists of raising awareness on the importance of sustainable solid waste management within and outside the community. At the start of the project, there were several activities to educate children, youth and adults from Sukunan on this topic. These included socialisation of residents, with discussions on waste management practices, painting, playing games (like competition for rubbish collection) and also singing. A motivational song 'Sukunan Bersemi' (Sukunan Blossom) and a poem 'Balada Seonggok Sampah' (Ballad of the Rubbish Heap) were created.

Due to the success of the community-driven waste management system, both Sukunan and the project got attention from local and national media (newspapers, radio stations and TV channels). Thus, the village opened its doors to visitors who wanted to learn about this community innovative initiative. In 2009, the community decided to turn Sukunan into an Eco-Edu Tourism Village for environmental education and awareness. Between 2009 and 2017 Sukunan received over 65,000 domestic and international visitors, with an average of 600 visits per month, including several school groups. The revenues from these activities, as well as the handcrafts and the selling of recycled material, go to a Village Fund. This fund is used to carry out activities within the village such as training, social activities, village development and construction of public facilities.

Over the 13 years of existence, the Sukunan project has been regularly monitored and evaluated by the community itself. The analysis includes number of families involved in the separation of waste, number of families involved in composting-making, number of women involved in craft-making, and outcomes such as waste collected per month and funding obtained per month. Monitoring and evaluation of waste sorting activities are conducted by three parties: the waste management team, garbage collectors and community members. The results are reported monthly and discussed at the village level.

The Sukunan project has received several awards. In November 2004, the Ministry of Environment, Government of Indonesia, evaluated several waste management projects and Sukunan was chosen as the winner because of the innovative character of the recycling activities and the integrated character of the community's involvement because the project included children and young residents, in addition to female and male adults. In 2006, it was selected as one of the best practices of community-based solid waste management in Indonesia. In 2010, the project was selected as 'the best of the best' in the 'Green and Clean' competition in Yogyakarta Province. In 2012, the Minister of Environment, Government of Indonesia, visited Sukunan and the village was declared 'Pioneer Village in the Response to Climate Change'.

The Eco-Edu Tourism Village project is a successful urban innovation. It is a peri-urban practice aiming at improving the conditions of the living environment (Meijer and Thaens 2016). It also represents a 'self-protecting innovation' (Glaeser 2011) as the same community has created the solution to solve their own environmental problem. Although there are other community waste management projects in Indonesia, such as the project in the Monkey Forest in Bali (MacRae and Rodic 2015), the Sukunan project represents a step further due to the holistic emphasis of the project. It is not only about managing waste but also creating change and environmental awareness within the local community. The success also lies in the collective effort demonstrated by Sukunan residents, with the majority of the villagers (both female and male and from a diverse age range) engaged in the project.

There are eight important factors explaining the success of this project, as follows: 1) Sukunan has a clear vision elaborated through community discussion and engagement to preserve the quality of the environment; 2) there is strong community commitment to carry out the project; 3) Sukunan established an organisation, which consists of trained personnel with strong commitment and solid internal bonds, to manage waste at the village level; 4) there are written local guidelines about waste management elaborated by the community and collectively agreed upon; 5) Sukunan residents participate in every stage of the decision-making process, empowering the community; 6) the project has been able to generate lucrative activities and home-based businesses for residents, which is considered a real economic benefit; 7) there is a solid and well-planned system for the recruitment process and involvement of community members in the project; and 8) Sukunan village is actively collaborating with other stakeholders, such as the government, private sector and universities to raise awareness on the value of community-based waste management initiatives.

Planning projects not only emphasise the importance of outcomes and end products but also the process to create change (Healey 1997). The Sukunan project shows the relevance of both process and outcomes. However, as discussed later, the project has not performed equally in the three areas of sustainability. The most significant outcomes are in relation to environmental benefits. Villagers have experienced the advantages of reducing waste, cleaning their living environment and reducing their health risks. Since there is no more burning of waste, the air is no longer polluted and residents can breathe fresh air. The village is cleaner and water flows are not blocked with waste. Illegal dumping sites have been eradicated from Sukunan. These are the main achievements of the project.

Nevertheless, social and economic achievements have also been important. The social results are about community participation, empowerment and engagement with the project. The majority of the residents (85 per cent) have become aware of the importance of solid waste management at the household level and involved in regularly conducting these practices at home. Community members have also acquired new skills, from learning how to mobilise and train their own community to sort waste and make compost, bricks and handcrafts. Additionally, there is a higher sense of cohesion within the community. The conflicts with farmers about garbage disposal that existed in the past have disappeared as waste is properly managed.

In regards to economic outcomes, residents are aware of the economic benefits of reusing, recycling and reducing waste. People on low wages have been able to increase their monthly incomes through selling compost and making handcrafts. Since they do it themselves, Sukunan residents do not have to pay for the waste management collection service by the local government (this is between IDR$10,000 and $50,000 per household per month – US$0.73 to $3.7 depending on waste volume), which represents a reduction in household expenses. The branding of Sukunan as an Eco-Edu Village and the prizes awarded to the community have turned the village in a tourist attraction. Visitors represent an additional source of income for Sukunan village (about IDR$54 million per year or US$4,000). In addition to these group benefits, the community member who works part time as waste collector gets an additional monthly income (IDR$400,000 or US$30).

As an Eco-Edu Tourism Village, Sukunan currently consists on seven centres of composting, 19 people involved in making handcrafts, one garbage depot (see Figure 7.3), an environmentally friendly house,[4] two farming areas, five waste-water treatment facilities where the water used by 150 households is treated, one centre for production of biogas from cow manure and a centre for *gamelan*.[5] The centres of composting are used for processing organic waste into fertiliser. The compost is used within the community and also sold to people outside the village. The waste bank of Sukunan is open every Sunday morning. Staff (community members) weigh the garbage deposited by each household, record information in a notebook and later sell the waste to wholesalers. Each waste-water treatment is used to process grey and black water from 30 households. The installations use biological waste-water treatment technology with a combination of anaerobic and aerobic systems. Effluent from the waste-water treatment is safely channelled into the rice field areas and rivers. A bio-digester of methane is used to treat the manure from 24 cows and the produced methane gas is used for cooking by local farmers. The centre of *gamelan* is used once a week by the community to practice *gamelan*. The Sukunan Tourism Village Team also offers the opportunity for visitors to learn how to play the *gamelan*. Two farming areas invite visitors to practice a variety of activities including ploughing fields using cows, land preparation, seeding, planting, weed cleaning and harvesting.

FIGURE 7.3 Waste depot in Sukunan

Photo by Sonia Roitman

Trade-offs

The project has faced challenges since its inception, as well as made some trade-offs over the implementation process. The first challenge was to realise that capacity building and raising awareness of the project within the community was not only important in relation to achieving the main outcome of the project (to have a cleaner village), but a requisite for collective action. The success of the project would depend on a collective effort. Therefore, it was a slow process of creating awareness and encouraging residents to participate that required the constant effort and strong commitment and enthusiasm from a few community members who acted as 'agents of change'. They also performed the role of 'environmental cadres' (Dhokhikah et al. 2015), who provide environmental counselling and guidance on waste management practices to the community. At the same time, the project leaders understood that it would not be possible to get all the community on board and they had to accept that not all residents would be participating in this project (85 per cent of the community participates).

The second challenge was to understand that this collective effort would also require negotiations within the community. Regular and open dialogue, both formally through organising focus group discussions and through informal meetings, was important. Creating this dialogue was also a slow process involving trust. It also led to some conflicts within the community. There was negative talk from several people in Sukunan community who felt they were not getting any financial benefits from the tourism activities within the village. This group argued only the Sukunan Eco-Village management team was getting economic benefits. This issue was solved by conducting a series of discussions which resulted in several agreements, such as opening opportunities for people outside the management team to be involved as tour guides and providing other services, such as snacks and beverages for visitors. A transparent schedule for task distribution was created so that many people could participate and benefit from tourism. Five per cent of the income from tourism is given to the village development fund for the establishment of public facilities, health services (especially for toddlers and senior members) and a community nursery for pre-kindergarten children. The distribution of the revenues from the sale of waste also created some community conflicts. Some residents felt they did not receive economic benefits because revenues were given to the Village Fund, not to each household. These residents argued that households that were involved in waste management were treated in the same way as those households that did not participate. The issue was solved by establishing a 'waste bank' open every Sunday where households could 'deposit' their sorted waste and receive some money for this. The revenues from selling the village waste were given to the waste depositor.

A third conflict was created with migrants moving to Sukunan over recent years. New houses have been built, mostly as boarding or rental houses. These 'temporary migrants' are less concerned about the waste management project and do not participate in sorting waste. Socialisation and persuasion have been conducted to solve this issue. The waste management team and the community leaders usually meet face-to-face with house tenants to explain the waste management system. If these households do not want to participate in the project, they are given an alternative service. Waste can be collected from their homes for a service fee that is between IDR$20,000 and $50,000 (US$1.4 to $3.5) per household per month depending on the waste volume.

Conflicts and issues have been solved through discussions and negotiations within the community. Most of the conflict resolution is done informally, such as hanging out at food stalls, because residents then feel more free to express their concerns, provide information and discuss solutions. The Sukunan management team is involved in this conflict resolution and there are regular meetings to discuss problems and brainstorm solutions. Community leaders are also involved in these meetings.

Sukunan is a poor community and was not able to cover the initial financial costs of the project. This was the third challenge as the community realised the need for external final support since the government and the local private sector were not able to provide this support. Hence, the community had to reach further out to get private funding to initiate the project. The project was funded by an individual family from Australia. The funding had several requirements. First, the funds should be given by an institution trusted by the donors. Even though it was from an individual family, it could not be handed over directly to Sukunan village. The funds were given through the director of ACICIS (Australian Consortium for In Country Indonesian Study) to the Sukunan waste management team. Second, Sukunan management team had to submit a project proposal to the donors with the explanation of how the waste management project would be implemented for the funding to be released. Third, funding expenditure was detailed and reported by the waste management team to the donors and Sukunan community. All this was done to create transparency and accountability.

The project has already implemented several practices to reuse and recycle most of the waste. There has been an intention to incorporate more advanced eco-practices, such as rain harvesting, reuse of water, reduction of electricity consumption, use of clay pots to keep fresh food as a replacement for refrigerators and urine recycling to process human urine into liquid fertiliser for plants. These practices have been successfully implemented in one house in the village. However, they require more complex adaptation practices and resources. It is not yet possible for most households to implement these practices. These eco-practice technologies aim to inspire visitors so that they could be replicated or modified in other settings. This has been an important trade-off since the project leaders became more aware of the need to implement practices incrementally. These more advanced practices will require more time and would probably not be implemented by many households in Sukunan.

Conclusion

A critical question for planning under the SDGs is 'How do we simultaneously protect the natural environment and reduce poverty and human injustice?' (Campbell 2016, 392). The Sukunan project is an innovative successful project that addresses these three areas of environmental concerns. Environmentally, it has been able to manage the majority of the waste produced in the village, through a bottom-up strategy collectively driven. Socially, the majority of the community participates in the project and makes decisions on project implementation. Economically, community members have been able to increase their incomes through the selling of recycled waste, compost and bricks and making handcrafts. An additional positive outcome has been that because of the success and notoriety that Sukunan has achieved, villagers feel very proud of their community and are willing to continue their efforts towards new innovation projects, such as more households treating their water and having their own organic gardens at home.

The Sukunan project is a very successful example of 3R solid waste management. By 2009 it had been replicated in 196 villages in Yogyakarta Province. The model was later developed into a waste bank model in 2009 and by 2015 had been implemented by 2,861 community groups in Indonesia (Ministry of Environment and Forests of the Republic of Indonesia 2015). In 2016, the model was replicated by community groups in Malaysia.

The success of the initiative depends on three critical aspects. The first one is the engagement and participation of the community and, as identified in similar projects, the community commitment to learning and constant effort (Bai et al. 2010). The second aspect is the existence of one or more leaders who can act as 'agents of change' or 'environmental cadres'. Additionally, community-driven projects usually face a financial obstacle to start the project and make it financially sustainable. In the case of Sukunan, the project would not have started without the external funding from international private donors (despite its small amount).

The 13 years of existence of the Sukunan project show its clear character of 'innovation', consisting of 'something proffered not as a one-off process but as something continuous' (Simone and Pieterse 2017, 18). The smooth implementation of the project only depends on the community, which is also the main beneficiary. The project does not currently rely on external supporters or facilitators, which contributes to the long life of this successful and innovative bottom-up solid waste management initiative. Despite the complex process of design and implementation, this project shows community-led initiatives can succeed and it offers a practical solution for Global South communities to take leadership on how to address SDGs 11 and 12 at the local scale.

Notes

1 The minimum wage in Yogyakarta is IDR$1,337,645 per month (US$90) (Decree of the Governor of Yogyakarta Province No. 235 / KEP / 2016, November 1, 2016).
2 It is believed that 6.3 billion tonnes of plastic have been produced worldwide up to 2015, with 79 per cent being accumulated in landfills or the natural environment, 12 per cent incinerated and 9 per cent is recycled (BBC News 2017).
3 The garbage collector is paid IDR$400,000 per month (one quarter of this salary comes from the sale of recycled waste, another quarter comes from the village tourism service, another quarter from the Village Fund and the final quarter from the environmental managers of Sukunan).

4 This house has a variety of simple green technologies such as water harvesting, natural home lighting and hydroponic vegetables.
5 The *gamelan* is a traditional Javanese musical instrument.

References

Bai, X., Roberts, B. and Chen, J. (2010) "Urban Sustainability experiments in Asia: Patterns and pathways", *Environmental Science and Policy* 13(4), 312–325.

Baye Alene, N. (2018) "The everyday politics of waste collection practice in Addis Ababa (2003–2009)", *Environment and Planning C: Politics and Space*, online version ahead of printing (http://journals.sagepub.com.ezproxy.library.uq.edu.au/doi/pdf/10.1177/2399654418757221).

BBC News (2017) "Seven charts that explain the plastic pollution problem", (www.bbc.com/news/science-environment-42264788) accessed 10 December 2017.

Bohm, R., Folz, D., Kinnaman, T. and Podolsky, M. (2010) "The costs of municipal waste and recycling programs", *Resources, Conservation and Recycling* 54, 864–871.

BPS (Badan Pusat Statistik) (2017) *Daerah Istimewa Yogyakarta Province in Figures*, Yogyakarta, Indonesia.

Campbell, S. D. (2016) "The planner's triangle revisited: Sustainability and the evolution of a planning ideal that can't stand still", *Journal of the American Planning Association* 82(4), 296–312.

Colon, M. and Fawcett, B. (2006) "Community-based household waste management: Lessons learnt from EXNORA's Zero waste management' scheme in two Indian cities", *Habitat International* 30, 916–931.

Dhokhikah, Y., Trihadiningrum, Y. and Sunaryo, S. (2015) "Community participation in household solid waste reduction in Surabaya, Indonesia", *Resources, Conservation and Recycling* 102, 153–162.

Fahmi, W. S. and Sutton, K. (2006) "Cairo's Zabaleen garbage recyclers: Multi-nationals' take over and state relocation plans", *Habitat International* 30, 809–837.

Friedman, J. (2013) "Memperkuat Lingkungan Kelembagaan untuk Manajemen Persampahan Perkotaan" [Strengthening the Institutional Environment for Urban Waste Management], *Journal Prakarsa Infrastruktur Indonesia* 15, 13–18.

Glaeser, E. (2011) *Triumph of the City*, Penguin, New York.

Government of Indonesia (2007) Law No. 26/2007 about Spatial Planning. Government of Indonesia, Jakarta.

Healey, P. (1997) *Collaborative Planning*, Macmillan, London.

Hoornweg, D. and Bhada-Tata, P. (2012) *What A Waste: A Global Review of Solid Waste Management*, World Bank, Washington.

MacRae, G. and Rodic, L. (2015) "The weak link in waste management in tropical Asia? Solid waste collection in Bali", *Habitat International* 50, 310–316.

Meijer, A. and Thaens, T. (2016) "Urban technological innovation: Developing and testing a sociotechnical framework for studying smart city projects", *Urban Affairs Review* 54(2), 363–387.

Ministry of Environment and Forests of the Republic of Indonesia (2015) *Inovasi Pengembangan Bank Sampah Sistem On-Line* [Innovation Development of Waste Bank On-Line System] (www.menlhk.go.id/berita-13-inovasi-pengembangan-bank-sampah-sistem-online-.html) accessed 8 December 2017.

Mursito, D., Sari, T. P. and Bramono, S. E. (2013) "Mengelola Sampah Perkotaan Indonesia, Sebuah Sudut Pandang Pemerintah" [Managing Urban Waste in Indonesia, A Governmental Viewpoint], *Jurnal Prakarsa Infrastruktur Indonesia* 15, 9–12.

Public Works of D.I. Yogyakarta Province (2016) *Pengelolaan Sampah* [Waste Management] (http://bappeda.jogjaprov.go.id/dataku/data_dasar/index/208-pengelolaan-sampah) accessed 2 December 2017.

Razak, N. (2010) "Partisipasi Masyarakat dalam Pengelolaan Sampah di Desa Sukunan, Sleman, Daerah Istimewa Yogyakarta" [Community Participation in Waste Management in Sukunan Village, Sleman, D.I. Yogyakarta Province] Unpublished Master thesis Department of Education and Environment, Universitas Sebelas Maret, Surakarta, Indonesia.

Rusqiyati, E. A. (2015) "Biaya pembuangan sampah TPA Piyungan diharapkan ditekan" [Piyungan Final Disposal Site costs are expected to be reduced] (https://jogja. antaranews.com/berita/329112/biaya-pembuangan-sampah-tpa-piyungan-diharapkan-ditekan) accessed 2 December 2017.

Simone, A. and Pieterse, E. (2017) *New Urban Worlds. Inhabiting Dissonant Times*, Polity, Cambridge.

Sukunan Village (2017) *Sukunan Village Population Book*, Sukunan, Sleman, D. I. Yogyakarta.

Tanaka, N. (2008) "Pengolahan Air Limbah Secara Komunal: Pengalaman Proyek Pusteklim di Yogyakarta" [Communal Waste Water Treatment: Experience of Pusteklim Project in Yogyakarta], in *Manual Teknologi Tepat Guna Pengolahan Air Limbah* [Manual of Appropriate Technology of Waste Water Treatment"], PUSTEKLIM Foundation Dian Desa, Yogyakarta.

UN (United Nations) (2015) *Transforming Our World: The 2030 Agenda for Sustainable Development, UN A/RES/70/1*, 21 October, United Nations, New York.

UN-Habitat (2016) *HIII Thematic Meeting on Informal Settlements. Pretoria*, March. UN-Habitat, Nairobi.

WHO (World Health Organization) (2009) *Environment and Health Risk: The Influence and Effects of Social Inequalities*, Report of expert group meeting, WHO Regional Office for Europe, Bonn, Germany.

Wijayanti, D. and Suryani, S. (2015) "Waste bank as community-based environmental governance: A lesson learned from Surabaya", *Procedia. Social and Behavioural Sciences,*= 184, 171–179.

Wilson, D., Velis, C. and Cheeseman, C. (2006) "Role of informal sector recycling in waste management in developing countries", *Habitat International* 30, 797–808.

8

SEVILLE, SPAIN

Improving cycling mobility in a city with no previous cycling culture

Manuel Calvo-Salazar and Ricardo Marqués

The City: Seville

Seville is located in southwestern Spain and has a population of approximately 700,000 inhabitants for the City and 1,400,000 for the whole Metropolitan Region, being the fourth largest city in Spain. It has an urban density of around 5,000 inhabitants/km² in the City and 3,500 inhabitants/km² in the urban areas of the Metropolitan Region. The per capita gross domestic product is EU€18,600 per inhabitant per year and the percentage of car-free households is over 20 for the whole Metropolitan Region. The percentage of university students is over 15 in the city and 8 in the whole Metropolitan Region. It is, therefore, a typical medium-sized southern European city with not a very high income and a relatively high level of motorization. Climate and topography (which is very flat) favors walking and cycling, except in summer, when temperatures get very high.

The local government is in charge of planning the infrastructure in the inner city through its Urbanism Department. Once built, this infrastructure is managed by the Mobility Department. The regional government, which is in charge of metropolitan mobility, finances and builds mobility and transport infrastructure at the metropolitan scale, whereas the small municipalities of the metropolitan ring around the inner city are in charge of mobility infrastructure inside these municipalities

The sustainability challenge: Reducing motorized traffic

Like many other southern European cities, Seville has experienced a sustained and significant increase in motorized traffic in the last four decades, as is shown in Figure 8.1.

However, people mobility, measured as the number of trips per day and per person, remained roughly constant in the same period. Thus, car trips increased at

Modal share

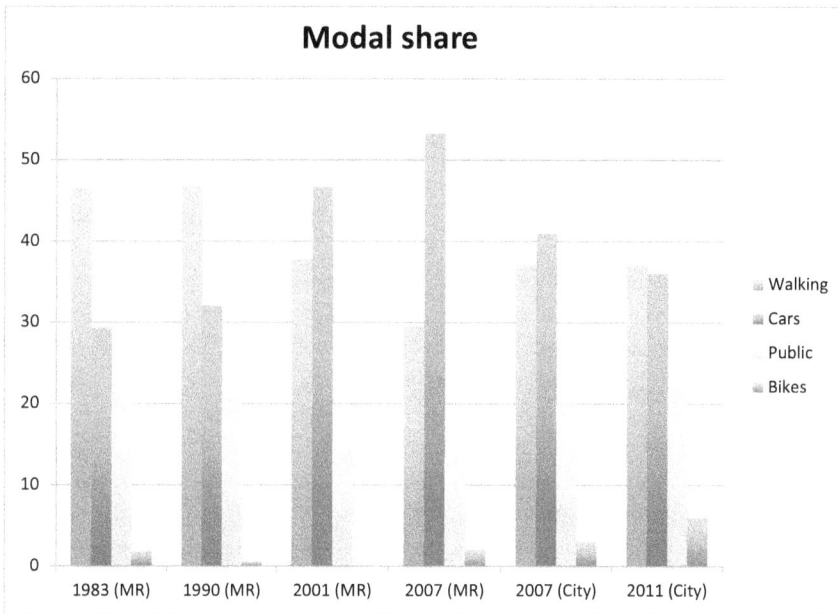

FIGURE 8.1   Evolution of the modal share in the Metropolitan Region and the City
of Seville

Source: Mobility surveys made by the regional government (Junta de Andalucía). Data for 2011 are
estimations (Marqués et al. 2014)

the expense of walking trips, despite the compact morphology of the city, which
is characterized by a high population density and a rather high mixture of urban
functions. This compact model is, however, in sharp contrast with the metropolitan ring, where low density and mono-functional built areas have been extensively
developed in the last four decades, which has also fostered an increased motorization. Thus, one of the main challenges faced by the City at the beginning of this
century was to shift the tendency towards an increasing motorization of mobility
and increase active mobility (Knoflacher 2007; Sanz-Alduán et al. 2014). Among
the initiatives aimed to solve such a problem, which is common to many other cities, the successful promotion of urban cycling through an ambitious infrastructure
program was the innovation in the Seville case.

The planning innovation

There are many cities in Europe with well-developed cycling mobility, so that
modal shares of cycling around 30 percent or even more are usual in northern and
central European cities (Pucher and Buehler 2008). However, as far as we know,
no other city above 500,000 inhabitants has reached the goal of multiplying by a
factor of six urban cycling mobility in just five years. This achievement, based on

the fast and complete building of a cycling network was probably the main innovation Seville applied to solve the aforementioned challenge. It is worth saying that, at this point, in Seville no cycle path was built as an isolated concept, but it was a complete network of cycle paths that was put in place, which meant that the network itself was the project.

The innovation process

Before 2006 Seville hosted a strong cycling movement and a comparatively high level of cycling (with regard to other Spanish cities) mainly supported by university people (the University of Seville had 70,000 people, or around a 10 per cent of the city population). The cyclists' main association – A Contramano – was founded in 1987 and during this period led many public demonstrations asking for bicycling infrastructure and made several proposals to the municipality for the effective implementation of such infrastructure. In 2003 a coalition formed by social democrats, leftists, and greens gained the city government and gave political support to these proposals. The planning of the current bicycle infrastructure began in 2006, following the guidelines contained in three main planning documents elaborated by the municipality:

- The "Urban City Master Plan" (Ayuntamiento de Sevilla 2006a), approved by the City Council in 2006. This document aimed to define the general urban criteria for the development of the City during the following decades. It introduced the concept and discussed the necessity of a bike network for the City, and defined a first "theoretical network" of cycle paths, which should be separated from motorized traffic as the main criterion for the network.
- The "Strategic Document for the Integration of the Bicycle in the Mobility System of Seville" (Ayuntamiento de Sevilla 2005) which informed the Urban City Master Plan and included a first proposal for a basic network of bike paths, a first proposal for the execution period of the public works (less than four years), a proposal for a bicycle parking program, and a first detailed proposal for bike path typologies along the network.
- The "City Bicycle Master Plan" (Ayuntamiento de Sevilla 2007). This document, after developing initial a research on the potential of cycling in the city, defined the basic network of cycle paths (77 km), as well as its basic construction criteria. It also defined the strategy for bicycle promotion in the historical center of the city and made some proposals for the improvement of links between bicycles and public transport, defined bicycle parking priorities and, for the first time, introduced the proposal of a bike sharing system for the city and defined the main criteria for its implementation. The City Bicycle Master Plan also included some "sectoral" programs for the promotion of the bicycle, such as "cycling to work" or "cycling to school". Finally, it determined the creation of a specific Bicycle Planning Department inside the Area of Urbanism of the municipality, which was in charge of the development of the plan.

A survey made within the framework of the aforementioned Bicycle Master Plan in 2006, before the network of cycle paths was built, evaluated the potential number of non-recreational cyclists in the city at around 89,000, more than 10 per cent of the city population (Ayuntamiento de Sevilla 2007). This survey also showed a high level of interest in the project among the population, providing strong political support for its effective realization.

Before 2006, there were some sparse and unconnected bike lanes, which after 2006 were restored and included in the network. From the total length of 120 km corresponding to the network in 2010, 77 km corresponded to the basic network defined in the Bike Master Plan, or what was called the first phase (Ayuntamiento de Sevilla 2007), and the remaining 43 km corresponded to the first extension of the network, or second phase. Further developments up to 180 km are described in Ayuntamiento de Sevilla (2017).

The starting point for the design of the basic network, or first phase, was a "theoretical network" (Ayuntamiento de Sevilla 2005), mainly based on the information collected during the elaboration of the City Master Plan. This theoretical network was strongly based on a former proposal of the local urban cyclists' association (A Contramano) which also played a crucial role in fostering the process, encouraging the city government to build the network of bicycle paths in the city, and persuading politicians of the suitability of the proposal. This first theoretical network connected major trip attractors of the city, such as main work and educational centres, main public transport hubs, and main relational spaces (such as squares, commercial streets, and green areas). This work was partially done during the elaboration of the City Master Plan, which included a parallel citizen participation process. During the discussions taking place in this process, which involved representatives of the cyclist and neighbor associations as well as experts and other stakeholders, some specific trip attractors – such as university centers – were positively biased because of their previous cycling tradition. A first version of this theoretical network was included in the City Master Plan, as a part of its determinations for city development.

The next step (Ayuntamiento de Sevilla 2005, 2007) was adjusting this theoretical network by optimizing the distances between the main trip attractors and the network, taking into account the space constraints and opportunities along the different streets. This adjustment was made in a qualitative way, on the basis of extensive fieldwork and discussions with stakeholders, and resulted in a first detailed proposal for the bike network. This first design also included proposals for bike path typologies along the network (Ayuntamiento de Sevilla 2005). It is worth mentioning that the aforementioned space constraints and opportunities usually favored the location of the bike paths along the main streets and avenues. This location was also favored by the analysis of the main trip attractors, which are usually located along such roads. Alternatively, locating bike paths along the main streets provides additional visibility to the network, a fact that was also taken into account in the design. As a result, bike paths were mainly located along the main streets and avenues of the city. Finally, after some final fine adjustments, more than 200 trip attractors were identified at a distance less than 300 m from the cycle network (Ayuntamiento de Sevilla 2006b).

In summary, the main characteristics of the resulting network were (Marqués et al. 2014, 2015a):

- Continuity and connectivity: the network was designed with the aim of connecting, through a continuum of bike paths, the main trip attractors and the main residential areas of the city.
- Cohesion and homogeneity: the design of the bike paths was very similar throughout the network, so that cyclists, and people in general, could easily follow and recognize them. This was achieved by using a uniformly colored (green) paving throughout the network, as well as a uniform morphology, which is described below.
- Directness and visibility: as we have already mentioned, the network followed the main streets and avenues of the city. Therefore it was quite visible. Moreover, as a general rule, detours and multiple street crossings were avoided.
- Comfort: the whole network was designed with the aim of providing comfortable bicycle riding for everyday cyclists, with parking facilities and a uniform and flat paving, without unexpected steps at intersections, etc.
- Quick construction: the whole basic network or first phase (77 km) was built in less than two years, during 2006 and 2007. So, in this first phase the aim was to have it under operation in a quick and complete way.

Of course, most of these characteristics – such as connectivity, cohesion, or directness – are recommended by many handbooks and are commonplace in many bike planning departments and publications (CROW 2007; Danish Cycling Embassy 2012). Other characteristics, however, can be considered as specific to the City of Seville, for instance, the quick building of the network or its extremely homogeneous design. The fast implementation of the first phase of the infrastructure was favored because of the simultaneous approval by the City Council of the City Master Plan that we have already mentioned. Therefore, the Bicycle City Master Plan could be considered as a specific development of the City Master Plan and benefited from the funding for that plan. The whole process was successfully carried out mainly by the creation of a specific administrative body in the Urbanism Department, the so-called Bike Office (Oficina de la Bicicleta), which managed the developments from planning to construction. This work methodology assured a homogeneous design as well as the financing of the infrastructure, despite several design teams participating.

Continuity, connectivity, and cohesion were considered as main criteria because the needs of non-recreational cyclists for transportation should be similar to those of other citizens. It was assumed that they would desire to move easily from one point to another in the city without leaving the cycle network except, perhaps, at the beginning or the end of their trips. Directness was considered important because bicycles are human-powered vehicles, and therefore it cannot be expected that cyclists make big detours along their trips. In fact, detour factors higher than 1.2 or 1.3 are considered excessive in most handbooks (CROW 2007). Visibility and homogeneity, as well as comfort, are important on their own, but they are even more important in cities without a previous utilitarian cycling tradition, because potential cyclists will only be persuaded to cycle if the cycling infrastructure is visible and comfortable, and can be easily interpreted. Finally, the quick building of the infrastructure was considered crucial because Seville did not have previous cycling infrastructures and so that the usefulness of the infrastructure would not be appreciated by the citizens until a basic network, connecting the main trip attractors, was completed.

Regarding the construction criteria of the bike paths, the planning determinations (Ayuntamiento de Sevilla 2007) and the physical constraints imposed by the previous city design generated these main design characteristics for the cycle paths (Marqués et al. 2015a):

- Separation: the whole cycle network was separated from motorized traffic by physical elements;
- Bi-directionality: most bike paths are bi-directional, with a width between 2.2 and 2.5 meters;
- Uniform pavement and signposting: bituminous pavement painted in green, with clear and uniform signposting, including specific traffic lights.
- Located between the motorized traffic zone (carriageway or parking lane) and the pedestrian area, following these criteria:
  - On the sidewalk, with a different pavement color and texture[1] and usually separated from the pedestrian zones by rows of trees or other discontinuous elements.
  - At the same level as the carriageway, but separated from it by bollards or other discontinuous physical barriers.
  - In case a parking lane was present in the street, the bike path is usually built at the same level as the sidewalk in order to facilitate the access to the parked cars for people with reduced mobility.
  - Intersections parallel to crosswalks, but separated from them and with specific signage.
- Built mainly over previous parking lanes and other spaces previously assigned to car traffic, so that the sidewalks were expanded, if necessary, in order to accommodate the bikeways.

An important aspect of the new cycling infrastructure of the City was the implementation of a public bike sharing system in July 2007, just a few months after the implementation of the first phase of the bikeways network. This system had 2,500 bicycles by the end of 2008, and after this date grew slowly to 2,600 bicycles in 260 stations throughout the City, with a separation of about 300 m between them. This system vas designed as an "individual public transport" (Marqués et al. 2015a)

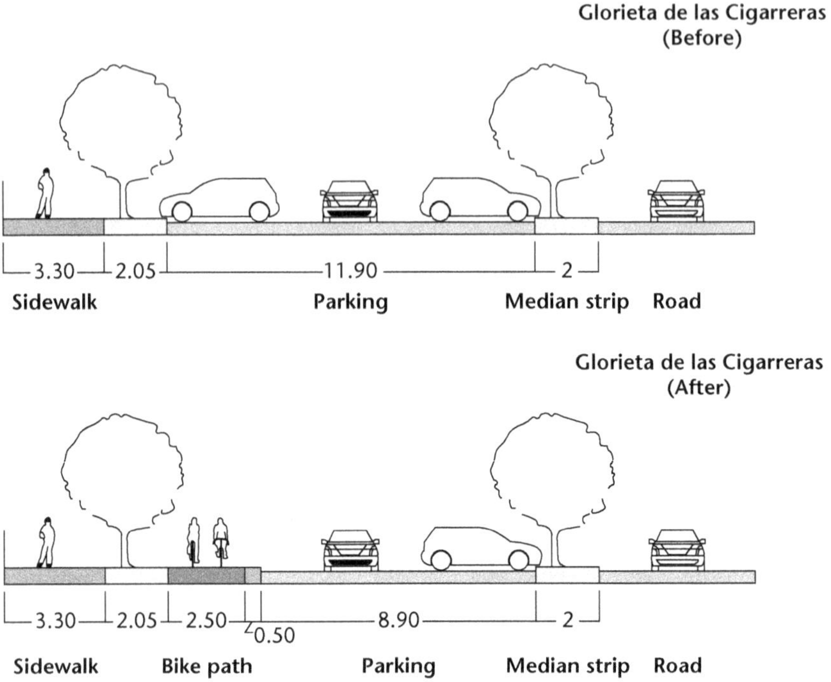

FIGURE 8.2 A typical section of a bicycle path on a street of Seville (before and after)

Source: Ayuntamiento de Sevilla

working 24 hours, 365 days per year and open to all people, including Seville citizens and visitors. Nowadays this system supports approximately 23 percent of the total cycling mobility in the city (Ayuntamiento de Sevilla 2017) and has a widespread acceptance among the population.

Finally it is worth mentioning that the building of the bicycle infrastructure was not an isolated process, but a part of a more complex strategy which included restrictions to non-resident private car traffic in the city center and the implementation of new public transport infrastructures, such as a tram along the main avenue of the city center and the first metro line connecting the City with some neighboring populations of the metropolitan ring (Castillo-Manzano et al. 2015).

Trade-offs

According to the above criteria, the whole of the network was separated from motorized traffic and almost all bike paths were bi-directional, with the aforementioned characteristics. Separation from motorized traffic, as well as continuity and perceived safety, were considered essential characteristics of the network design (Ayuntamiento de Sevilla 2006a, 2007).

Discussions with civil society were also very important and were carried out through the creation of a Bike Civic Commission (Comisión Cívica de la Bicicleta) where transparency and commitment were put in place. In such an institution all cycling stakeholders were represented. This methodology allowed that every step taken by the administrative bodies regarding cycling development and promotion had been previously discussed with civil society and so they had been properly endorsed.

Of course problems arose once it was clear that the bike paths were taking space from car traffic and parking. The media, shoppers, and some residents from some neighborhoods complained about the project. In such cases, especially with residents, a discussion took place in order to negotiate final design solutions. The result was generally an agreement. It is also worth noting again the important role played by the Bike Office in close work with the Urbanism Department, which managed the political issues.

In this context, there were many issues that were under discussion. Bi-directionality was controversial, because in many cities with a long cycling tradition, like Amsterdam or Copenhagen, bike paths are usually mono-directional. It is beyond the scope of this chapter to develop a complete discussion about the advantages and disadvantages of both designs. Bi-directional bike paths were preferred in Seville mainly because they save space and are much cheaper and faster to build (a mono-directional cycle path must be at least 1.50 m in width, whereas a bi-directional cycle path can be 2.5 m width (Ayuntamiento de Sevilla 2007; CROW 2007)). Moreover, in a city with a limited cycling tradition, it can be expected that mono-directional bike paths would be used as bi-directional ones, as actually happened in Seville with the few previously existing mono-directional bike paths, thus creating conflict once bicycle traffic increased. Another reason that favors bi-directional bicycle paths in the case of Seville is that it was considered important, at least in a first stage, to accommodate all the cyclists in the same corridor, in order to increase their visibility and therefore, their safety.

Placing bike paths at the sidewalk level also was controversial because of the conflicts generated with pedestrians (Marqués et al. 2015a). However, when a parking lane persisted in the street (the most common case), this design provided for easy access to the parked cars from the sidewalk while keeping the separation between the bike path and the motorized traffic. Also, the perceived safety of the network is higher for the cyclists. This last effect was considered important at the first stages of the planning. However, the new Bicycle Master Plan (see below) considers building protected bike lanes at the carriageway level as the basic design for the future, in order to avoid the aforementioned conflicts with pedestrians once the first stages of cycling development have been implemented (Ayuntamiento de Sevilla 2017).

It is worth mentioning that, even if many cycle paths are presently placed on the sidewalks, most of them were built on previous parking lanes (Hernández-Herrador and Marqués 2017). This loss of car parking places was also controversial, but helped to reduce car traffic. Figure 8.3 shows a photograph of a typical bike path on a busy street, illustrating this change of land use. The overall changes in land use caused by

FIGURE 8.3 Typical bike path along a main avenue of the city

Note: The row of cobblestones next to the alignment of trees marks the end of the old sidewalk. The cycle path is built on a former parking lane.

Source: Photograph by R. Marqués

the implementation of the basic network of bike paths have been recently quantified by Hernández-Herrador and Marqués (2017). This analysis shows that the whole process implied a net change of use for 14.7 ha of the city surface that are now dedicated to bicycle paths. Of these 14.7 ha, 53 per cent were obtained from the carriageway, 38 percent from unpaved areas, and 9 per cent from pedestrian areas.

In the following years, the cycle network continued its expansion, with the same criteria and the same construction characteristics, until it reached the present length of 180 km (including recreational and pedestrian-shared paths), which makes up 12 percent of the total road length in the city (Ayuntamiento de Sevilla 2017). It now makes a homogeneous and continuous network connecting the most important trip attractors and residential areas through the main streets and avenues of the city, with a typical cell size of about 500 m. It is determined that new urban developments should be connected to this network by new bike paths and be built following similar criteria of design. It is also determined that the network should be completed by other complementary developments, such as parking infrastructure, at a smaller scale inside residential and industrial areas (Ayuntamiento de Sevilla 2007), although this last determination is being applied very unevenly in practice. In any case, it can be said that cycling from one point to another in the city can now be done mainly by following the cycle network with safety and comfort.

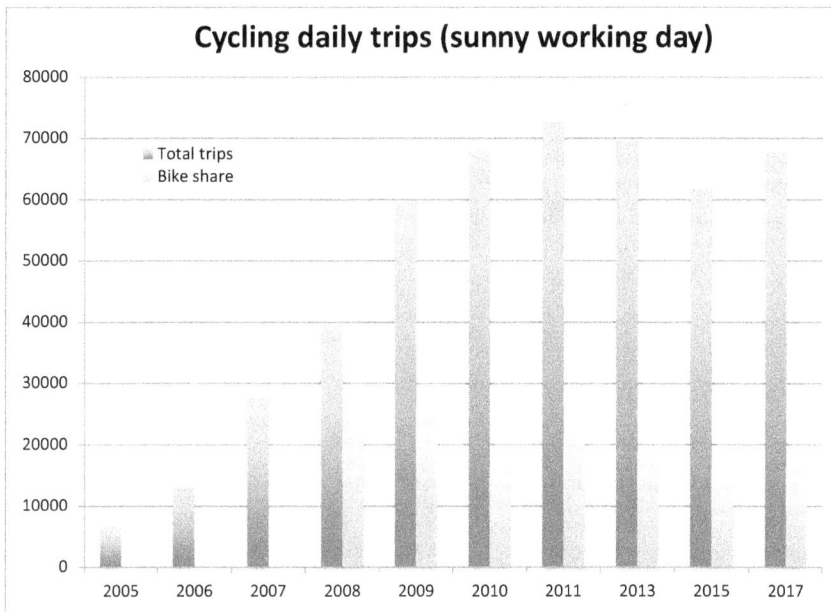

FIGURE 8.4 Evolution of cycling mobility in the City of Seville

Note: Data were taken yearly in a sunny working day of the month of November, when usually the peak of cycling mobility takes place.

Source: Marqués et al. (2015a), Ayuntamiento de Sevilla (2017) and authors' own research

The impact of the cycling infrastructure on cycling mobility was dramatic. As shown in Figure 8.4, the number of bicycle trips on a daily basis increased from approximately 13,000 to a peak of 72,000 in November 2011 (Marqués et al. 2015a), due not only to the implementation of such infrastructure but also to the restrictions on car traffic in the city centre imposed to non-residents (Marqués et al. 2015a).

The impact of the network of bicycle paths on cycling safety was also positive. A recent study (Marqués and Hernández-Herrador 2017) has reported a sudden drop in the risks associated with cycling (defined as the number of crashes between bicycles and motor vehicles per million bicycle trips) in the City after the implementation of the network of bicycle paths. According to this analysis, the risk dropped by a half after 2006, not just as a consequence of the increase in the length of the network but also as a consequence of the creation of the network itself, and in close connection with the "safety in numbers" effect (Marqués and Hernández-Herrador 2017).

Besides this meaningful increase in cycling safety, other positive effects such as important economic and health benefits also came from the implementation of the network. Health benefits in the City have been evaluated as 24 deaths avoided each year (Marqués et al. 2015a), while a recent cost-benefit analysis of the infrastructure (Brey et al. 2017) reported a current net value of €557 million, and an internal rate of return of over 130 percent, mainly in terms of sustainability, health, and leisure time. It is also remarkable that the only economic activities that have been growing during the economic crisis that started in 2008 are those related to cycling, such as retail, leisure, and tourist services.

Even though there are no systematic studies about the social impact of the growth of cycling mobility, it has been obvious that there have been several social groups that have been direct beneficiaries. Indeed, the youth have discovered a more reliable and affordable means of transportation. The same applies to those with very low wages, since cars are still an expensive means of transportation for many. In general, the city itself has benefited because the quality of public space is much better, especially in some areas where the presence of bikes has helped to reduce car speed.

New bicycle trips are now almost equal with individual motorized modes (cars and motorbikes), public transport modes, and walking trips (Marqués et al. 2015a), and more than 50 per cent of these trips were to work or to school, which reveals a predominantly non-recreational use of the bicycle (Marqués et al. 2015a), while women's share of bicycle trips increased from 25 per cent in 2006 to 35 percent in 2017 (Ayuntamiento de Sevilla 2017). Regarding age, 39 per cent of cyclists are below 29 while only 33 percent of the city population is below 29,

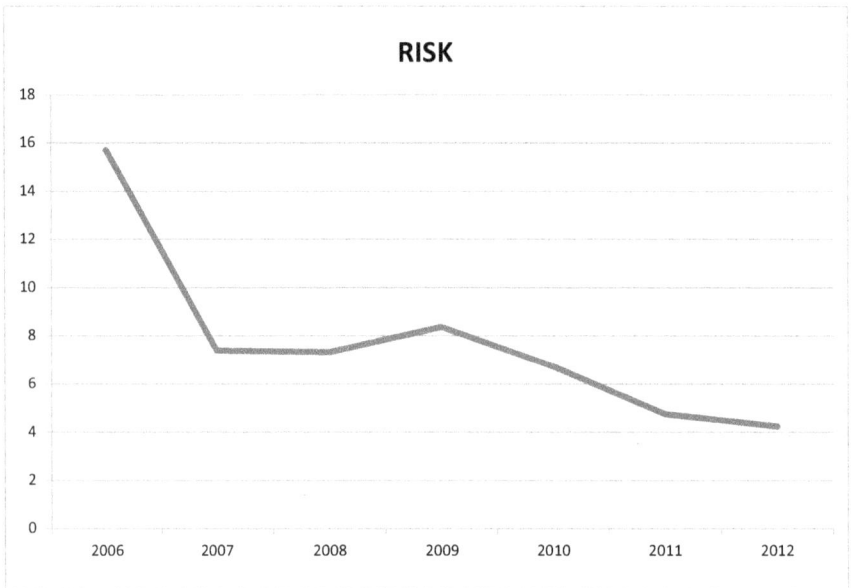

FIGURE 8.5 Evolution of the risk of cycling (bicycle–motor vehicle crashes per million bicycle trips) in the City of Seville in the years after the building of the cycle network

Source: Marqués and Hernández-Herrador (2017)

which shows that the urban cyclists are, on average, younger that the population of the City, but also that citizens of all ages are represented among the cyclists (Marqués et al. 2015a). Only for the elderly does cycling sharply decline (only 4 per cent of cyclist are above 64). In summary, after the period 2007–2011, cycling became part of the urban culture of the City, and the bicycle became a useful means of transport for many people regardless their age or gender (Huerta and Hernández-Ramírez 2015).

After 2011, cycling mobility started to decline (see Figure 8.4), reaching an absolute minimum of 62,000 trips/day in 2015. These changes occurred in parallel to a sudden change in the mobility policies of the City government, which changed after the local elections of 2011. The newly elected City government returned to more car-oriented policies and suddenly stopped pro-bike policies and traffic restrictions in the city center. This negative dynamic persisted until 2016. Since then, cycling has been on the rise again, in parallel with the return to more pro-bike policies by the present City government, whose political orientation changed again after the local elections of 2015. Right now, around 68.000 cycling trips take place in the city on a typical working day (Ayuntamiento de Sevilla 2017).

These changes in cycling mobility, which correlate with changes in the orientation of the mobility policies of the successive City governments, suggest a strong influence of such mobility policies on cycling. They also suggest that a continuous political effort must be applied in order to maintain and improve the positive cycling mobility trends of the City. Thus, during 2017, the municipality was working on developing a new master plan (Ayuntamiento de Sevilla 2017), with the objective of doubling the percentage and number of daily cycling trips, up to 115,000, which would make up 15 per cent of total vehicular trips.

Some measures proposed by the new master plan are:

- To develop soft cycling promotion policies, including education and social interventions in order to promote cycling at homes, jobs, schools, and urban services.
- To substantially improve safe bicycle parking, increasing the number of racks in the streets and the infrastructures for safe parking into residential buildings, workplaces, and schools, as well as in the main public transport hubs, in order to promote high levels of intermodality.
- To improve the quality of the cycling network and its maintenance.
- To improve local traffic regulations in order to promote cycling.

As a result, the municipality is working on improving the quality of the network by trying to detect and solve risk points, especially at intersections, and widening the parts of the network that channel higher bicycle traffic. The objective of this enhancement is to increase the capacity of the infrastructure in order to attract more people to cycling through the improvement of safety and convenience.

A key aspect of the new master plan is to develop specific policies in order to provide safe bicycle parking at origins (homes) and destinations (jobs and schools, urban infrastructures, shopping areas and malls) of cycling trips. Several programs on these issues are now being developed.

Improving bicycle and public transport links is also a key aspect of the new bicycle master plan. In fact, improving such links through the provision of safe bicycle parking at the main metropolitan public transport hubs has been identified as a key factor for improving the population coverage of the public transit system, mainly in the metropolitan ring of the city (Marqués et al. 2015b). Taking into account these results, the municipality of Seville is now working on improving safe parking at the main metropolitan transit nodes within the city. Unfortunately, similar intermodality policies are not being developed by the municipalities surrounding Seville or by the regional government, which is responsible for mobility policies at the metropolitan scale. Therefore, the overall impact of such policies, developed only at the scale of the municipality of Seville, will be by no means complete.

Conclusion

The experience of Seville has been, and still is, remarkable in many aspects. The cultural and social dimensions of the process were significant and implied substantial changes in land use and mobility in the city. Some conclusions that can be extracted from this process are identified in this final section.

The case study suggests that it is possible to achieve a fast growth in urban cycling mobility if active policies and fast development of cycling infrastructure are provided. This suggests that low levels of cycling mobility in cities are not a necessary consequence of the lack of urban cycling culture or an ideological issue. If there is enough political will and a commitment to develop the appropriate infrastructures, people will start cycling as long as the bicycle becomes a comfortable, effective and convenient means of transport.

To achieve this objective, strong local cyclists' associations and the commitment of these associations to the objective of promoting cycling in the City are important factors for success.

In the case of Seville, besides the rapid growth of the numbers of everyday cyclists, two additional effects were significant: first, an integration of the bicycle in the urban culture and the landscape of the city, which allowed for a widespread and transversal use of this vehicle by the whole population, quite irrespective of age and/or gender; and second, a positive effect on the development of similar initiatives in other cities, as the experience of Seville was disseminated and became known and analyzed in other cities.

In order to achieve further development of cycling mobility, Seville has to persist in applying active policies on cycling promotion and motorized traffic restrictions and to find the way for integrating the municipalities of the metropolitan ring in these policies. This need for active policies is expressed in the fact that during the period between 2011 and 2015, when the local government was not supporting pro-bike policies, cycling trips dropped from 72,000 to 62,000 trips/day, and since the new local government took office in 2015 and returned to more pro-bike policies, cycling trips have risen again (up to 68,000 trips/day).

Our analysis suggests that, in order to fulfill the main objectives of the new bicycle master plan, further measures and policies other than cycling promotion are needed. Indeed, the evolution of city cycling mobility, which peaked in 2011 in parallel with the implementation of car traffic restrictions, suggests that such restrictions play a central role in cycling promotion. Restrictions on car movement and parking have to be applied. These policies have to tackle especially the present pattern of urban space distribution and assignments that mean the car is the leading actor in up to 70 to 80 per cent of this space. This distribution must be changed to maybe no more than 50 per cent in avenues and no prominence at all in residential areas where cars have to be considered more as guests.

Our analysis shows that cycling mobility in the City of Seville covers a fraction of the mobility demand that is typically urban (mid-length trips) and, in cooperation with mass transit, also metropolitan. The analysis of the impact of the cycling infrastructure of Seville on bicycle traffic safety has shown that this impact has been highly positive in terms of reduction of the risks associated with cycling (traffic accidents per million bicycle trips). The economic analysis of the experience of Seville shows that cycling infrastructure is a very efficient public investment. The return rate of these investments is very positive and higher than many other transport infrastructures, mainly in terms of sustainability, health, and leisure time.

From a practical point of view, there is no reason to think that the experience that is being carried out in Seville could not be transferred to any other city. What is clear is that a strong political will is essential. What is more important is that this commitment has to be maintained, at least for enough time for cycling to be understood, and so the bike is used as a comfortable and convenient means of transport. For this purpose, building the infrastructure in a fast and homogenous way is essential. Once a complete infrastructure is in place, and people start using it, the process is unstoppable. Cycling culture is created very rapidly. The fact that Seville did not previously have such a cycling culture demonstrates that this condition is not a prerequisite for success as many people, and also some scholars, suggest.

These conclusions suggest that cycling promotion and usage is open to many cities around the world and that a sudden change in mobility patterns is possible if sufficient political will drives the process. The role of civil society is also important. In fact, we have learnt that the energy that derives from the civil movement helps this political will to be harnessed at the beginning of the process and then maintained. Indeed, maintaining this political will is, in the end, the only way of maintaining the wave for long enough to be surfed by everyone else.

Note

1 These design characteristics later became enforced by law by the regional government (Junta de Andalucía) for any bike path built along pedestrian areas or sidewalks.

References

Ayuntamiento de Sevilla (2005) *Bases y estrategias para la integración de la bicicleta en la movilidad urbana de Sevilla*. Gerencia de Urbanismo, Ayuntamiento de Sevilla.

Ayuntamiento de Sevilla (2006a) *Plan General de Ordenación Urbana (Plan de Sevilla), Sevilla*; www.sevilla.org/plandesevilla/adef/contenido.html

Ayuntamiento de Sevilla (2006b) *Vías ciclistas de Sevilla: dotaciones, equipamientos y espacios libres*. Gerencia de Urbanismo, Ayuntamiento de Sevilla.

Ayuntamiento de Sevilla (2007) *Plan de la Bicicleta de Sevilla 2007–2010*. www.sevilla.org/sevillaenbici/plandirector/PlanBiciSevilla.html

Ayuntamiento de Sevilla (2017) *Programa de la Bicicleta Sevilla 2020*. www.urbanismosevilla.org/areas/sostenibilidad-innovacion/sevilla-en-bici/programa-bicicleta-2016-2020

Brey, R., Castillo-Manzano, J. I., Castro-Nuño, M., Lopez-Valpuesta, L., Marchena-Gomez, M., and Sanchez-Braza, A. (2017) Is the widespread use of urban land for cycling promotion policies cost effective? A cost-benefit analysis of the case of Seville. *Land Use Policy*, 63 (April), 130–139.

Castillo-Manzano, J. I., López-Valpuesta, L., and Marchena-Gómez, M. (2015) Seville: A city with two souls. *Cities*, 42, 142–151.

CROW (2007) *Design Manual for Bicycle Traffic*. CROW, the Netherlands.

Danish Cycling Embassy (2012) *Collection of Cycle Concepts 2012*. www.cycling-embassy.dk/wp-content/uploads/2013/12/Collection-of-Cycle-Concepts-2012.pdf

Hernández-Herrador, V. and Marqués, R. (2017) El impacto del "carril-bici" de Sevilla sobre el espacio urbano de la ciudad: un análisis preliminar. *Hábitat y Sociedad*, 10, 181–202.

Huerta, Elena and Hernández-Ramírez, Macarena (2015) *Etnografía de la bicicleta en Sevilla*. DAPS Informes y documentos de trabajo. https://rio.upo.es/xmlui/handle/10433/1425

Knoflacher, H. (2007) Success and failures in urban transport planning in Europe: understanding the transport system. *Sadhana*, 32(4), 293–307.

Marqués, R. and Hernández-Herrador, V. (2017). On the effect of networks of cycle-tracks on the risk of cycling. The case of Seville. *Accident Analysis & Prevention*, 102, 102 (May) 181–190,

Marqués, R., Hernández-Herrador, V., and Calvo-Salazar, M. (2014) Sevilla: A successful experience of bicycle promotion in a Mediterranean context. *WIT Transactions on Ecology and the Environment*, 191, 769–781.

Marqués, R., Hernández-Herrador, V., Calvo-Salazar, M., and García-Cebrián, J. A. (2015a) How infrastructure can promote cycling in cities: Lessons from Seville. *Research in Transportation Economics*, 43, 31–44

Marqués, R., Hernández-Herrador, V., Calvo-Salazar, M., Herrera-Sánchez, J., and López-Peña, M. (2015b) When cycle paths are not enough: Seville's bicycle-PT project. *Urban Transport*, 21(146), 79–91.

Pucher, J. and Buehler, R. (2008) Making cycling irresistible: Lessons from the Netherlands, Denmark and Germany. *Transport Reviews*, 28(4), 495–528.

Sanz-Alduán, A, Vega, P., and Mateos, M. (2014) *Cuentas ecológicas del transporte en España*. Libros en Acción.

9

SEOUL, SOUTH KOREA

Dismantling a highway – Cheonggyecheon Restoration Project

Sun-Young Rieh and Ji-in Chang

Introduction

Cities are shifting from car-oriented roads to pedestrian-friendly streets, and the dismantling of highways associated with this trend is bringing about dramatic changes in the urban landscape and beyond. The Cheonggyecheon Restoration Project (CRP) represents a paradigm shift in metropolitan urban planning and urban governance in South Korea. It provides a template for green urban regeneration, public conflict management and economic revitalization through restoration of nature. The reason that CRP resonates with cities worldwide is because it shows that sustainable urban solutions can successfully address and mitigate a wide range of economic, social and environmental urban problems. Since the CRP was completed in 2005, Seoul has become the champion of 'green growth', transforming itself into a more people-oriented city with an emphasis on welfare and equity.

Cheonggyecheon (清溪川), meaning 'crystal clear stream' in Chinese characters, runs east–west through the historic centre of Seoul from surrounding mountains and valleys. It is a 10.92 km long urban waterway that has been closely linked to everyday life in Seoul, South Korea's capital city. However, with the aftermath of the Korean War and Seoul's industrialization starting in the mid-1950s, it served as a sewage canal, lined with squatter housing and illegal industrial businesses. The smell of the sewer and the squalid conditions needed urgent measures. Seoul city government decided to cover the 2.4 km length of the stream and build a road over it. This construction lasted from 1958 to 1961. Soon after, between 1967 and 1971, a 5.7 km-long elevated arterial highway was added with the aim of accommodating heavy traffic and alleviating a road shortage. By the 1980s, however, the congestion and pollution of the Cheonggyecheon area epitomized the worst aspects of uncontrolled urbanization. Until the early 2000s, Cheonggyecheon elevated highway was for through-traffic, while the streets underneath were used as service spaces for parking and loading. Over time, the Cheonggyecheon area experienced industrial decline while the environment deteriorated further as air and noise pollution became ever more serious.

Through the dismantling of the elevated highway and the restoration of the stream, the Cheonggyecheon area was revitalized and led to the redevelopment of the entire district. The CRP was a 'flagship' stream restoration project with an estimated cost of US$366 million (Green 2011). Recovering the stream and bringing nature back into the heart of the city resulted in an increase of pedestrians. More offices, housing and commercial facilities were built, which increased the number of residents. The environmental change brought about by the restoration of Cheonggyecheon transformed the historic urban centre into a people-centred eco-friendly environment.

The city: Seoul

Seoul is the capital city of the Republic of Korea, covering an area of 605.21 km^2 with a population of 9.86 million people in 2015. Seoul Capital Area, comprised of Seoul, Incheon Metropolitan City and Gyeonggi Province, is the third largest metropolitan area in the world by population. With a population of 25.6 million people in 2015, more than half of Korea's total population lives in the Seoul Capital Area. As Korea's capital for more than 600 years, Seoul occupies a central position in Korea's culture, economy, politics and education. In the global hierarchy, Seoul aims to be a global financial centre in East Asia; Seoul is ranked sixth in the 2017 Global Power City Index (Mori Memorial Foundation 2017). Seoul's population density in 2017 was 16,700 persons/km^2, the highest among cities within the member nations of the Organisation for Economic Co-operation and Development (OECD). It is four times greater than Paris (3,550 persons/km^2), more than three times greater than London (5,100 persons/km^2) and Tokyo (4,750 persons/km^2) (City Mayors 2017). Seoul has a polycentric structure with one dominant centre and five sub-centres.

Seoul is composed of 25 districts, or '*gu*', with local autonomy since 1995. Seoul has an elected mayor, the head of the Seoul Metropolitan Government (hereafter referred to as SMG) which administered a budget of almost $28 billion in 2017. Seoul raises almost 84 per cent of its budget through local taxation, which reduces its reliance on the central government and gives it significant autonomy. Seoul's legal status was elevated in 1962 to be under the jurisdiction of the prime minister from being under the jurisdiction of the minister of home affairs. Consequently, the mayor of Seoul is involved in both the formation of policies at the national level and the implementation at the local level.

The urban sustainability challenge: Ageing infrastructure and the decline of the old city centre

In the 1950s after the Korean War, with the influx of migrants from the countryside to Seoul, the banks of Cheonggyecheon were lined with squatter housing.

The natural stream became heavily polluted, effectively becoming an open sewer subject to seasonal flooding during the rainy seasons. Beginning in 1959, the informal houses were cleared and the stream was covered over with concrete. In the 1970s, a 16 metre-wide elevated highway was constructed over Cheonggyecheon as part of an arterial road linking the eastern parts with the western parts of Seoul. By the 1980s, Cheonggyecheon had become a source of congestion and pollution. Seoul's historic city centre began to decline as businesses moved to Gangnam area, south of the Han River, which divides Seoul into northern and southern parts.

One of the most urgent reasons for the restoration of the Cheonggyecheon area was imminent safety risks. Even as early as 1991-1992, the Korean Society of Civil Engineers reported corrosion of the elevated highway's steel frame and structural flaws in the upper plate. Two kilometres of the elevated highway received repairs. In 1997, the use of the overpass was restricted indefinitely to light vehicles only. According to safety evaluation by the Korean Society of Civil Engineering in 2000, the highway was a safety hazard due to deficiencies in its structural strength. In 2000-2001, another examination showed continued serious deterioration of the upper slab and concrete beams. Demolition and a total reconstruction of the highway was recommended with an estimated cost of $95 million over three years. The high cost of reconstruction made the alternative plan of sustainable development more viable.

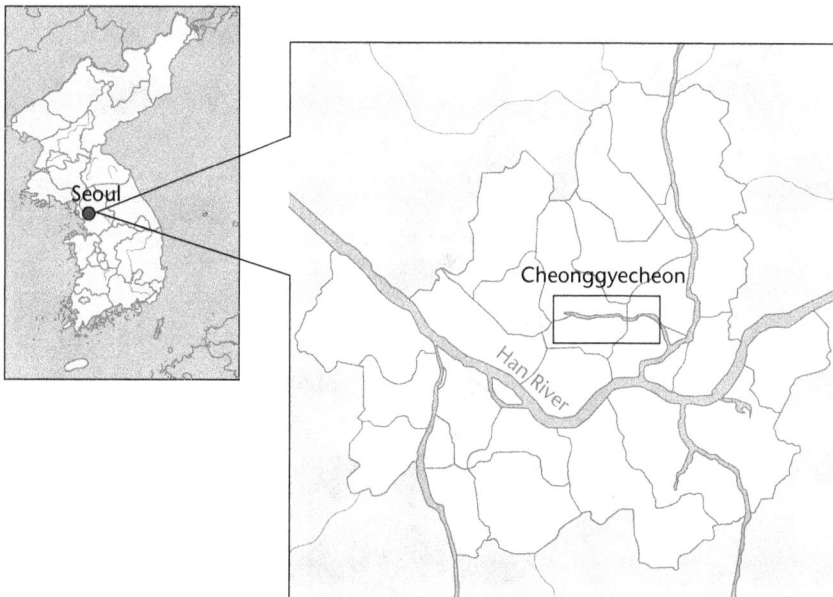

FIGURE 9.1 Location of Cheonggyecheon

Source: the authors

FIGURE 9.2 Cheonggyecheon highway and overpass before the restoration

Source: Courtesy of Cheong Gye Cheon Museum

FIGURE 9.3 Cheonggyecheon Restoration Project under construction in 2004

Source: Courtesy of Cheong Gye Cheon Museum

Korea has adopted the term 'sustainable development' over 'sustainability' in its policies, indicating its priorities. Sustainability began to be actively discussed in Korea as a result of Agenda 21, adopted by more than 178 governments at the United Nations Conference on Environment and Development (UNCED) held in Rio de Janeiro, Brazil (Kim 2007). It can be argued that 'sustainable development' was embraced in Korea after it gained currency in the West in keeping with its 'catch-up' practice as a developmental state.

The terms 'sustainability' and 'sustainable development' appear similar, but they lead to distinctively different approaches. Whereas 'sustainability' puts the focus on the natural environment regarding its restoration and/or protection by changes in human behaviour and goals, 'sustainable development' incorporates economic growth and human development into the equation (Evans 1997; Robinson 2004). This is why 'sustainability' is espoused by environmentalists and non-governmental organizations (NGOs), and 'sustainable development' is referenced by governments and the private sector. Having wholly adopted and incorporated the definition 'sustainable development' in the Brundtland Report into planning policies, the discourse in Korea embraces the 'three Es' of environment, economy and equity in planning (Berke 2002; Berke and Manta-Conroy 2000; Campbell 1996).

There were environmental challenges in CRP related to air quality, urban heat island, waste disposal and biodiversity. From 1958 to 1978, Cheonggyecheon was covered over by a road about 6 km long and 50-80 m wide and a 16 m wide overpass. These roads were used mainly for through traffic (62.5 per cent) by more than 168,000 cars per day. The highway built over Cheonggyecheon was a major source of pollution and traffic congestion in the downtown area. Air pollution containing carcinogens exceeded the environmental air quality standard, leading to respiratory diseases for those who lived or worked in the area.

Improvement of air quality and mitigation of the urban heat island phenomenon are the most tangible effects in environmental changes after the restoration. The most prominent is the ambient temperature change after the stream's restoration. This phenomenon is due to the formation of a wind path associated with the demolition of the highway, and in particular the reduction of the ambient temperature through the water's surface evaporation. In addition, heat island reduction was also brought about by the overall temperature drop through the introduction of greenery. These changes in temperature around Cheonggyecheon contributed to the pedestrian-friendly environment, thus increasing the number of people using the stream's side walkway during the hot summer season. According to SMG statistics, the average temperature of the site was lowered by 1.3 degrees Celsius after CRP (Seoul Solution 2015). In the case of air quality change, the concentration of NO_2 decreased from 51.0 to 46.0 parts per billion (PPB) by more than 10 per cent and the fine dust decreased from 71.0 $\mu m/ m^3$ to 60.0$\mu m/ m^3$ by 16 per cent (Kim and Song 2015).

The congested roads before restoration also created heat islands, which were reported to be more than 2.2 degrees Celsius above the surrounding area. This was due to a chain reaction with the reflectivity (albedo) on the highway. The reduction of the road, the introduction of greenery, the absorption of the heat of vaporiza-tion through the water surface, and the formation of the wind tunnel through the removal of the highway structure all contributed greatly to the improvement of the urban heat island phenomenon.

Another environmental challenge to be dealt with was the management of construction waste. The dismantling of the city highway and the restoration of the submerged stream inevitably led to the problem of waste disposal. In case of CRP, 1,140,000 tons of waste was produced by the dismantling of the highway and removal of the covered road over the stream. It is reported that 96 per cent of total waste resulting from the construction was reused (Seoul Solution 2014). Concrete waste was used to construct the promenade on both sides of the stream and to construct the walls on both sides, which not only reduced the cost and effort in regard to discharging wastes, but also drastically reduced the need for new materials.

The greatest significance of the restoration of the Cheonggyecheon stream was in the recovery of an environmentally friendly urban space by creating an ecological waterway in the city centre. With the restoration of Cheonggyecheon, it has been reported that the number of flora and fauna has increased from 98 to 626 species (Seoul Solution 2015). It is clear that it has succeeded in creating an ecological environment to some extent; it is possible to observe plants and animals in the centre of the city and to find migratory birds. Even though it is an artificial stream, various species live in it, creating the possibility of an ecological passage connecting the city and the Han River. The success of the Cheonggyecheon stream as an ecological stream has been a catalyst for urban regeneration elsewhere in Korea.

Second, there were economic and equity challenges related to the economic decline of the old city centre, restoring the balance between north and south of Seoul, the transition to a pedestrian-friendly city and transit system reform. Cheonggyecheon is located in the historic centre of the capital city. The historic central business district (CBD) was in decline. Between 1991 and 2000, the CBD experienced a population decline of 50,000 people accompanied with a loss of 80,000 jobs. The number of businesses in the CBD decreased about 23 per cent from 100,200 to 77,000, even though the overall number of business increased 24.6 per cent for the whole of Seoul. The number of workers during the same period declined 5.4 per cent to 400,000 persons. The CRP was an opportunity to increase economic and environmental competitiveness as well as balanced devel-opment in relation to the booming Gangnam area, situated to the south of Han River which runs through metropolitan Seoul.

Cheonggyecheon's restoration has become a catalyst for economic revitali-zation of the old CBD. Recreational and cultural amenities along the restored stream have spurred economic development. It has become a desirable loca-tion, generating new jobs and economic activities. The number of businesses in the area increased by 3.5 per cent from 2002 to 2003, which is even before the completion of the CRP. This was double the increase in the number of businesses for the rest of downtown Seoul. Over the same time period, the number of workers in Cheonggyecheon area increased by 0.8 per cent compared to a decrease of 2.6 per cent in other parts of downtown Seoul (Landscape Architecture Foundation 2011).

Although the Cheonggyecheon restoration does not include the upstream part of the stream and has been criticized as an imperfect reconstruction that recycles the water from Han River, the powerful impact on the CRP on urban regeneration and played a critical role in diluting the social stratification phenomenon between north and south of the stream. The restoration of existing historical relics and the installation of new bridges contribute to the linking of 5.8 km of urban stream and its surroundings into an ecological place where history, culture and nature coexist. The north and south separated by the stream were connected smoothly through the restoration. It also became a catalyst for the balance between the more modern and prosperous Gangnam, south of Han River, and the historic and declining Gangbuk, north of Han River.

The dismantling of the Cheonggye highway and the restoration of the stream meant a reduction of road surface. In order to prevent traffic congestion, the entire Seoul public transportation system, which included bus routes and connections to subway stations, was reorganized in advance. As the roads on either side of the stream could not accommodate the existing traffic volume, they were converted into one-way streets to keep the traffic along the stream

FIGURE 9.4 Cheonggyecheon after restoration

Source: The author, Sun-Young Rieh

FIGURE 9.5   Cheonggyecheon at night

Source: Courtesy of Cheong Gye Cheon Museum

to a minimum flow. The public transportation system was expanded in adjacent blocks and supported by improvements to subway and bus lines to strengthen the transportation spine. The relocation of bus stops made access to public transportation more convenient.

The urban regeneration project led to extensive traffic system reform. The CRP needed to address issues of traffic flow during and after the construction period. Measurements in 2008 show that the overall vehicle speed in the CBD increased after the demolition of the elevated highway in 2002. By 2008 there were 170,000 less car trips per day. The use of public transportation increased by 4.3 per cent for the subway and 1.4 per cent for buses. Taking Seoul as a whole, the traffic system reform resulted in an increase in passengers of 15.1 per cent for buses and 3.3 per cent for the subway (Kim et al. 2009).

The provision of dense public transportation, service and consumption oriented-functions along the street, attractive night life and parking spaces around business and commercial facilities located at the upper stream encourage the commuting patterns of the workers towards a more pedestrian-friendly direction. Also, as there is a lack of space to accommodate the leisure activities of citizens in the CBD, Cheonggyecheon has become an important historical, cultural and recreational space. An entrance square of almost 2,200 m^2 made it possible to attract popular annual events, such as the Cheonggyecheon Birdwatch Festival, Luminarie Event, Hi Seoul Festival, etc. In particular, the night lighting connected with the water space plays a key role in regenerating the city. It has helped to restore and revitalize the area's cultural legacy through the excavation of the waterbed and through cultural festivals. In particular, the different designs of the 22 bridges connecting the surrounding areas help people to become orientated at the lower stream level.

The planning innovation

The CRP has shown the positive impact of bringing nature back into the urban centre. It has changed the way sustainable development is perceived in both Korea and around the world. Its economic impact in terms of inducing further development and job creation along the stream has led to a paradigm shift in terms of creating value for ecological environments, improved quality of life and recreational amenities in city centres. As a case in point, Ryu and Kwon (2016) found that between the period of 2006 and 2015, there were 49 new building developments, 140 extensive renovations and numerous mergers of small building plots allowing for greater high-rise developments within a 3 km area encompassing 12 blocks from Cheonggyecheon entrance square. It has galvanized other cities in both Korea and abroad to recover their own streams. For example, the CRP has provided a template for urban greening of downtown areas as well as for regeneration through linear urban parks. As such, CRP has contributed to Seoul's balanced development. It has provided a turning point for revitalization of the central city, rediscovery of the historical city and a shift of power around the Han River from Gangnam (southern part) to Gangbuk (northern part) in Seoul. In addition, Cheonggyecheon is visited by millions of visitors, and it has become a landmark civic space in central Seoul. Surveys show that the satisfaction levels of all users and visitors of the restored Cheonggyecheon area are high. Cheonggyecheon is a linear public recreation space visited by 70 million domestic and foreign visitors each year. Kim et al. (2009) calculated that with 64,000 foreign visitors daily spending on average 1.5 million won, their daily contribution to the local economy amounted to $1.9 million (2.1 billion won). Extensive pedestrian networks have encouraged walking instead of driving. Had the elevated highway remained, the maintenance cost would have been $90 million and would have required three years of repairs. While the total cost of maintenance and repairs would have amounted to about $260 million, the Cheonggyecheon restoration project has resulted in an estimated $1.98 billion of new capital investment after the redevelopment (Hwang 2007).

Innovation process

Because of Cheonggyecheon's historical relevance, the restoration project was not seen as just another urban planning project; it was a symbolic task and the entire nation was eager to revive its historical and natural heritage from the start (Lah 2011). The project was completed on October 2005. When the CRP was announced, citizens were able to visualize future benefits, aided by positive media coverage of the CRP and the project's likely impact. In this aspect, the media played a crucial role in garnering public support. Despite its publicity and public support amounting to 86 per cent of those surveyed, the project lacked public confidence. Most Seoulites wanted to see the stream opened up, but doubted the project's feasibility. Mayor Lee persuaded opinion leaders, formed a city restoration task force, advertised the CRP on street banners and subway bulletins, made media appearances and actively defended the project against opposition (Lah 2011).

The idea of revitalization of Cheonggyecheon first emerged among engineering professors in the late 1990s. A research forum was formed and scholarly symposiums were held. The then-mayoral candidate, Lee, Myung-bak, adopted the idea as a campaign issue and in 2002, as the newly elected mayor of Seoul, proposed restoring Cheonggyecheon. He believed that the CRP would be beneficial to urban development, the economy and the greening of the urban centre. The envisioned CRP drastically shifted the city's urban policy paradigm from development to sustainability (Lah 2011).

In 2003, Mayor Lee began the restoration of the Cheonggyecheon stream by demolishing the elevated highway and removing the concrete road covering the natural stream. CRP Headquarters was the project's major engine. Seoul Institute (previously called Seoul Development Institute) was the brain. SMG established the Cheonggyecheon Restoration Centre to act as a focus for research and planning. Seoul Institute was the city's research institute that supplied scholars and researchers specializing in urban planning, environment, economics and public administration. The Seoul Institute not only prepared the master plan for the architectural restoration, but also was charged with anticipating possible difficulties in the implementation process, such as conflicts with citizens.

The Cheonggyecheon Citizens' Committee helped to gauge public opinion, communicating the project's goal through information sessions and conveying concerns. The Cheonggyecheon Citizens' Committee consisted of academics, journalists, pastors, and professionals (Lah 2011). The CRP project took two years. As a public-private partnership turnkey project, the CRP was divided into three segments, each under the charge of a different consortium of major Korean construction companies.

The scope of the project was strategically defined in consideration of the budget and the difficulty of execution in the private realm. SMG decided to limit the restoration project to public land along the Cheonggyecheon stream. Strong opposition by local businesses and retailers called for a short construction period to minimize economic downtime. Cheonggyecheon restoration led by the public sector was followed by urban regeneration of the surrounding private sector areas. Benefits were shared by both public and private sectors, as the synergy effects, such as increased numbers of pedestrians staying or passing, created much higher property values.

Under the strong leadership of Mayor Lee, SMG established the CRP Headquarters (project management and coordination), Cheonggyecheon Restoration Research Corps (urban planning) and the Cheonggyecheon Citizens' Committee (consensus building). In other words, the CRP Headquarters was the project's main engine, whose responsibilities included conflict mitigation and negotiations with local businesses.

The communication channel was the Cheonggyecheon Citizens' Committee, consisting of 127 members, which provided a link between the public and the city by facilitating communication and exchange of ideas. It was divided into a main committee and six sub-committees focused on a particular subject area: natural environment, history and culture, urban planning, transportation, construction safety and citizen communication. Local residents, business owners and NGOs were not included as members, leading to doubts about the adequacy of the committee's functioning truly representatively as the citizens' voice. However, the three implementation organizations were strongly interlinked: the headquarters was in charge of execution, the Seoul Institute was the brain and the committee functioned, albeit with restrictions, as the eyes and ears.

Trade-offs

Social trade-offs are evident in the relocation of the existing tenant merchants and street vendors. Local stakeholders could be categorized as property owners, tenant merchants or street vendors. Property owners were beneficiaries in that the value of the land increased after restoration. Tenant merchants were dissatisfied with the CRP because the rents would increase and force them to relocate. The industries of the CRP were interdependent. Merchants opposed the project, but eventually gave in because of political pressure (Lah 2011). Street vendors were most dissatisfied, and after heated negotiations, were provided with temporary facilities in Dongdaemun Sports Stadium. Of the 3,265 merchants surveyed, 95.75 per cent opposed the CRP (Lah 2011).

Cheonggyecheon Business Area Defenders United, consisting of seven shopping centre merchants and 21 separate organizations, were loosely organized, but managed to hold several rallies against the restoration project. The Clothes Stores Association was the most vociferous among them. Their demands included compensation for lost sales and relocation of their businesses. In addition to demands for direct talks with the mayor, they also sought help from the city council, political parties and the media. SMG's strategy was to provide indirect benefits to the affected merchants (for example, low interest loans) by offering relocation to new shopping areas in exchange of land, and by giving priority to Cheonggyecheon-area businesses in terms of government's purchase of commodities. It would have been impossible to finance the project if local businesses affected by the restoration had been directly compensated. The city's economic and political manoeuvring broke down the organizations. In the end, their opposition was ultimately over-ruled by public opinion.

The Seoul Institute conducted research, opinion surveys and statistical analysis of industrial networks to back up the expectations regarding the completion of the project with data and helped build support by citizens based on evidence. SMG officials travelled at least 3,000 times to the Cheonggyecheon area to talk to stakeholders. Their effort built public trust and showed the commitment of the city government. The Citizens' Committee was a major communication channel, but SMG conducted public hearings, operated an on-site complaints centre, and undertook regular policy conferences, official site visits and meetings with the mayor (Lah 2011). SMG also tried to minimize inconveniences of doing business by providing interim public parking connected by a Cheonggyecheon area shuttle. The SMG provided financial assistance to improve city markets and made public loans available at a low interest rate. It built an alternative shopping centre to which Cheonggyecheon merchants could move if desired.

Another socio-economic trade-off for revitalization of the CRP is loss of affordable commercial rents due to increased price of property and gentrification. Erection of high-rise buildings along both sides of the stream led to greater density of population. Land prices increased, which resulted in the relocation of small businesses. Traffic volume became greater. Higher density and rising real estate prices on the banks of the stream have led to gentrification of the area, with low-income residents being replaced by affluent newcomers who can afford the higher rents and land prices. Property values adjacent to the stream increased by 300 per cent. The price of land increased by 30 to 50 per cent for properties within 50 metres of Cheonggyecheon. This increase was double the rate of property cost increases for other parts of Seoul (Kim et al. 2009).

The environmental trade-off revolves around the question of CRP's authenticity as a waterway. The artificially created stream consumes a large quantity of energy to keep the appropriate volume of water flowing. As it is essentially a dry stream, the restored Cheonggyecheon needs to be pumped with 120,000 tons of water daily from the Han River. As it is not a natural stream, it requires daily maintenance, such as keeping the stream clear of moss and other contaminants, and maintaining the water quality and natural landscaping, among other things. In the long term, plans are to be implemented to connect to natural water sources. Maintenance and water pumping costs were reported as being more than 7 billion won ($8 million) per year. That includes cleaning the bottom of the stream, controlling water quality, and gardening and landscaping. In addition, ecological authenticity needs to be restored by completing the recovery of the entire stream. Only 5.8 km of the total 10.9 km length have been uncovered; the other half of the stream is still used as a sewer. The rest of the stream needs to be restored to enable fish and other wildlife to live on their own without human intervention.

Conclusion

Despite numerous shortcomings, the CRP is internationally recognized as a successful urban greening development in a metropolitan context. It made possible the return of natural features into the city by dismantling an overpass to allow more sunlight and air flow. In terms of replicability, it shows first, the importance of leadership and policy that is dependent on strong local power base, and the importance of high local autonomy in terms of revenue and resources. The CRP was government funded and could offer stakeholders incentives and financial support for inconveniences incurred during construction. It provided alternative transport and parking spaces. Second, it engaged the stakeholders in consenting to an environmentally sustainable model of development through effective communication. This required numerous consultations and the effective use of media to create positive public support. Last, and most importantly, the CRP showed that a polluted and congested urban space could be turned around to become a sustainable green public place, leading to enhanced quality of life and economic revitalization of the entire city. Through this process, it has altered the value system that had been driven by property market objectives and physical manifestations of development.

Three major lessons can be drawn from the CRP case study. First, strong political leadership bolstered by financial resources was essential to undertake the urban development project. Second, an evidence-based approach supported by strong institutional research capacity was instrumental in inspiring trust and cooperation among stakeholders. Finally, the importance of citizen involvement cannot be overemphasized.

In terms of sustainability, the adoption of the concept of sustainable development has brought great changes in planning for Korea. It has been instrumental in the inclusion of many different groups into the process of planning. It has weakened the monopoly held by select occupational groups in policy implementation. It has contributed to a greater transparency and public involvement in the planning system. Often the endeavour to pursue environmentally friendly and equitable economic development is essentially hindered by a contradiction that pits sustainability on one side and development on another. This conflict is inherent in the tension between environmental protection and economic growth, equity and efficiency, jobs and environment (Berke and Manta-Conroy 2000; Robinson 2004). Every city faces this dilemma, but the case of the CRP shows that sustainable urban development, even if it seems to privilege physical planning, can be achieved.

References

Berke P. (2002) "Does sustainable development offer a new direction for planning? Challenges for the twenty-first century". *Journal of Planning Literature*, 17(1), 21–36

Berke P. and Manta-Conroy M. (2000) "Are we planning for sustainable development? An evaluation of 30 comprehensive plans". *Journal of the American Planning Association*, 66(1), 21–33

Campbell S. (1996) "Green cities, growing cities, just cities? Urban planning and the contradictions of sustainable development". *Journal of the American Planning Association*, 62(3), 296–312

City Mayors (2017) www.citymayors.com/statistics/largest-cities-density-125.html. Accessed 13 January 2018

Evans B. (1997) "From town planning to environmental planning" in Blowers A. and Evans B. eds, *Town Planning into the 21st Century*, Routledge, London, 1–14

Green C. (2011) Case study brief – The restoration of the river Cheonggyecheon, Seoul, SWITCH: Managing Water for the City of the Future (www.switchurbanwater.eu/out puts/pdfs/W6-1_GEN_DEM_D6.1.6_Case_study_-_Seoul.pdf) accessed 15 December 2017

Hwang K. (2007) "Cheonggyecheon restoration and downtown revitalization". Paper for the Conference on When Creative Industries Crossover with Cities, Hong Kong Institute of Planners (http://www.hkip.org.hk/CI/paper/Prof.%20Hwang.pdf) Accessed 14 January 2018

Kim B. (2007) "Seoulsi cheonggyecheonbogwonsaeob-e daehan dosiyuhyeonglon-jeog bunseog (Urban typological analysis of the Cheonggye Stream Restoration Project in Seoul)". *Journal of Korea Planners Association*, 53 111–130

Kim H. S., Koh T.G. and Kwon K. W. (2009) "The Cheonggyecheon (Stream) Restoration Project – Effects of the restoration work", Cheonggyecheon Management Team, Seoul Metropolitan Facilities Management Corporation

Kim K. and Song J. (2015) "Cheonggyecheon bogwonsaeob-i dosiyeolseomhyeonsang-e michineun yeonghyang (The effect of the Cheonggyecheon restoration project on the mitigation of urban heat island)". *Journal of Korea Planners Association*, 50 (4), 139–154

Lah T. J. (2011) "The huge success of the Cheonggyecheon Restoration Project: What is left?", in Holzer M., Kong D. and Bromberg D. eds, *Citizen Participation: Innovative and Alternative Modes for Engaging Citizens: Cases from the United States and South Korea.* Rutgers, Newark, 97–117

Landscape Architecture Foundation (2011) https://lafoundation.org/myos/my-uploads /2011/12/15/cheonggycheonrestorationmethodology.pdf. Accessed 2 January 2018

Mori Memorial Foundation (2017) www.mori-m-foundation.or.jp/english/ius2/gpci2/. Accessed 2 January 2018

Robinson, J. (2004) "Commentary: Squaring the circle? Some thoughts on the idea of sustainable development". *Ecological Economics*, 48, 369–384

Ryu, C. and Kwon Y. (2016) "How do mega projects alter the city to be more sustainable? Spatial changes following the Seoul Cheonggyecheon Restoration Project in South Korea". *Sustainability*, 8(11), 1–7

Seoul Solution (2015) http://seoulsolution.kr/node/3250. Accessed 2 January 2018

Seoul Solution (2014) https://www.seoulsolution.kr/en/content/seoul-urban-regeneration-cheonggyecheon-restoration-and-downtown-revitalization. Accessed 2 January 2018

10

LOS ANGELES, UNITED STATES

Adaptive re-use of buildings to regenerate the inner-city

Amanda Napoli, Sébastien Darchen and Donald Spivack

Introduction

Adaptive re-use can be considered as a tool for urban regeneration. The City of Los Angeles has experienced relative success in applying this tool to regenerate a declining downtown into a thriving city-centre, thus contradicting the very idea of a postmodern dysfunctional city. A major positive aspect of this regeneration initiative from the City's perspective is the low cost of the operation, as the process relies mainly on the role of developers to take advantage of the Adaptive Re-use Ordinance (ARO) in transforming vacant office buildings into marketable and often high-priced residential lofts. The regeneration has its shortcomings, but over-all it has had a tremendous impact in shifting perceptions of a decaying downtown area into a desirable place to live and invest. Urban regeneration seldom delivers sustainability for all three components (economy, social, and the environment) but in theory, it has the potential to do so. From a social sustainability perspective, the implementation of the ARO has led to gentrification and contributed to the expulsion of low-income groups; however, it has been a powerful tool in initiating a shift from a sprawling metropolis towards a denser, walkable, and more sustainable urban environment through reinvestment in the Downtown. The analysis is based on qualitative data collected through semi-structured interviews as well as an analysis of planning regulations associated to the ARO. We also mapped the location of adaptive re-use projects in the downtown area.

The City: Los Angeles

The urban area of Los Angeles makes up the most densely populated large urban area in the United States and the nineteenth largest built-up urban area in the world at 6,000 people per square mile, with a land area of 2,432 square miles (Demographia 2016).

TABLE 10.1 1960-2010 Historical US Census population of Los Angeles County and the
City of Los Angeles

	April 1, 1960	April 1, 1970	April 1, 1980	April 1, 1990	April 1, 2000	April 1, 2010
Los Angeles County	6,038,771	7,041,980	7,477,238	8,863,164	9,519,338	9,818,605
City of Los Angeles	2,479,015	2,811,801	2,968,579	3,485,398	3,694,742	3,792,621

Source: California Department of Finance (2013)

It is primarily due to the higher density suburbs of Los Angeles that this area is considered the densest in the US, in stark comparison with its lower core density.

As shown in Table 10.1, Los Angeles has had a steadily increasing population and is the second most populated city in the US after New York City (City of Los Angeles 2015).

Through the 1970s and 1980s, Los Angeles was staunchly viewed as a polycentric city with many jobs focused in several prominent employment centres (Gordon et al. 1986). This polycentric urban form was also a result of planning: in the 1970s, the City adopted a "Centres Plan" that consciously accepted and promoted a polycentric city, identifying 56 high intensity centres (18 of them outside city limits) and providing for upzoning in these areas and downzoning in the interstices (City of Los Angeles Department of City Planning 1970).

Jobs in Downtown LA represented less than 5 per cent of the region's employment during the mid-1990s, and continued to decrease in the years following. This lack of employment within Downtown LA was in large part due to low and moderate population densities of the Southern California region and the employment lure of the suburban "edge cities" in the region (City of Los Angeles 2015).

The City, incorporated as an American city in 1850, is a charter city, meaning that it is generally independent of State Legislature in matters concerning municipal affairs (City of Los Angeles 2015). Los Angeles is governed by an elected mayor and a 15-member council. City Council has authority to adopt zoning and other land-use controls, in addition to several other powers, and can override a veto by the mayor with a two-thirds vote (City of Los Angeles 2015). In 1999, the City introduced neighbourhood councils in an attempt to decentralize planning reviews and introduce reform (Whittemore 2012). It is only more recently from the late 1990s onwards, that Los Angeles has begun producing an environment with high-density residential developments and the use of urban infill. Downtown LA's population essentially tripled from 18,700 in 1999 to 57,797 in 2014, with an increase of 22,150 residential units during this period (Downtown Center Business Improvement District 2014). Joh et al. (2015) identify the greater LA area as being in a crucial transitory stage where it is becoming a multi-modal metropolis primarily due to increased investment in transit and downtown revitalization projects, which includes adaptive re-use projects. Initiatives such as the Mobility Plan 2035, with an accompanying Mobility Hub Program, the Transit Neighborhood Plans Program, and LA River Revitalization Master Plan Project exemplify just a few of the recent planning strategies the City is investing in to help revitalization (City of Los Angeles Department of City Planning 2017).

The urban sustainability challenge: Residential flight

The sustainability challenge was a decaying downtown and urban sprawl following the residential flight towards the suburbs. This challenge was the outcome of two phenomena: suburbanization and urban renewal initiatives that led to the displacement of residents, thus accelerating the decline of the Downtown.

With suburbanization and prioritization of the personal vehicle and extensive freeway network of Los Angeles after the Second World War, Downtown LA began its decline. Industrial neighbourhoods Downtown during this time, such as what later became the Arts District, began to encounter challenges with the evolution and shifting of industrial practices. The Arts District today is a regionally significant historical and cultural neighbourhood on the east side of Downtown Los Angeles. It contains a collection of former industrial buildings converted over the last 20 years to lofts, creative office space, and a lively arts and restaurant scene (Banuelos 2014; Darchen 2016).

With the emergence of the trucking industry as the primary means of goods movement and the growth of mechanized manufacturing came a need for both increased arterial-type roads and single-storey industrial sites. This demand pushed many corporations to purchase larger land parcels in the newer cities outside of Downtown LA, including Vernon and the City of Commerce. Vacancies in Downtown factories and warehouses increased dramatically as a result and decline in the inner city followed, as both workers and work places relocated to the urban fringes (Los Angeles Conservancy 1999). These outlying new suburbs amassed the majority, first, of manufacturing facilities and later, corporate headquarters and offices, facilitated by the growing freeway network and suburbanization of the LA workforce. With the multibillion-dollar Federal Highway Program and growth of suburban housing through the Federal Housing Administration, cities such as Los Angeles became viewed as "obstructions to the flow of traffic," and "junkyards" of substandard housing (Berman 1982).

Bunker Hill, a former middle-class residential neighbourhood situated within Downtown LA, housed many working-class immigrants in tenement structures through the 1940s. This late nineteenth century neighbourhood, once a fashionable residential community overlooking Downtown, had deteriorated dramatically over the years. The once stately mansions had been converted to apartments and rooming houses and were by then viewed as an "eyesore" by city officials who categorized the area as "blighted," with a particular emphasis on its racial diversity (Avila 2004, 58). This resulted in a redevelopment scheme to demolish much of the growth that had occurred in the neighbourhood over the prior hundred years. The Bunker Hill Urban Renewal Project, approved in 1959, instigated the demolition of 7,310 housing units, disrupted the established social networks of Bunker Hill's residents, and contributed to Downtown LA's decline in several ways (Marks 2004). The displacement of residents in Bunker Hill – outside of the extremely low-income residents of Skid Row, the only sizable resident population Downtown and its largest base of consumers – led to the closure of many Downtown stores, concurrent with relocation of retail outlets, warehouses, and offices to the suburbs. The redevelopment, with a focus on signature office, hotel, and higher-income residential towers built on newly defined "superblocks" with large setbacks, wide boulevards, vast parking garages, and minimal street-level pedestrian-oriented activity, symbolized the new landscapes and spatial reorganization that would occur in Downtown LA and many other American cities during this time period. The geographic isolation of Bunker Hill – some 90 feet above the rest of Downtown LA – contributed further to the decline of the rest of the Downtown core. Even as the outmigration of business to the suburbs slowed as several major firms chose to move to new quarters on Bunker Hill, the abandonment of the rest of Downtown LA continued.

During the early 1960s and 1970s, Downtown LA's Skid Row neighbourhood lost approximately 7,500 of its 15,000 very low-income housing units due to demolitions of residential hotels in the area which did not meet fire and safety codes. To conform to the codes, owners were mandated to either repair or demolish these units and most chose the demolition route as it was the least costly (Community Redevelopment Agency of the City of Los Angeles 1998). This substantial loss of housing units exemplifies the urban renewal and clearance-type tactics that erased large swaths of American cities during this time period. The demolition and clearing of land in cities such as Los Angeles not only contributed to a loss of housing units but also displaced many residents.

The decline of Downtown LA following the post-war period sets the stage for exploration of adaptive re-use tactics that would later stimulate urban regeneration.

The planning innovation: Adaptive re-use

A conspicuous solution in such a context is the adaptive re-use of vacant buildings; in the case of Los Angeles, those primarily consisted of office buildings. Adaptive re-use can be defined as "a process to ameliorate the financial, environmental and social performance of buildings . . . that changes a disused or ineffective item into a new item that can be used for a different purpose" (Bullen and Love 2010, 215). More simply, this is a process whereby a structurally sound older building is repositioned for economically practical new uses (Austin et al. 1988). Adaptive re-use of buildings in city centres contributes to sustainability through the mitigation of CO^2 emissions and the re-deployment of existing structures, thus reducing growth at the urban rim, along with using already-in-place infrastructural capacity such as streets and water and sewer systems. The built environment has a prominent role to play in the debate on sustainable development and climate change as it demands 40 per cent of global resources and generates a high amount of waste (Langston 2008).

As pressure increases for residential accommodation in major North American cities, including Los Angeles, a variety of unique sustainable options to encourage

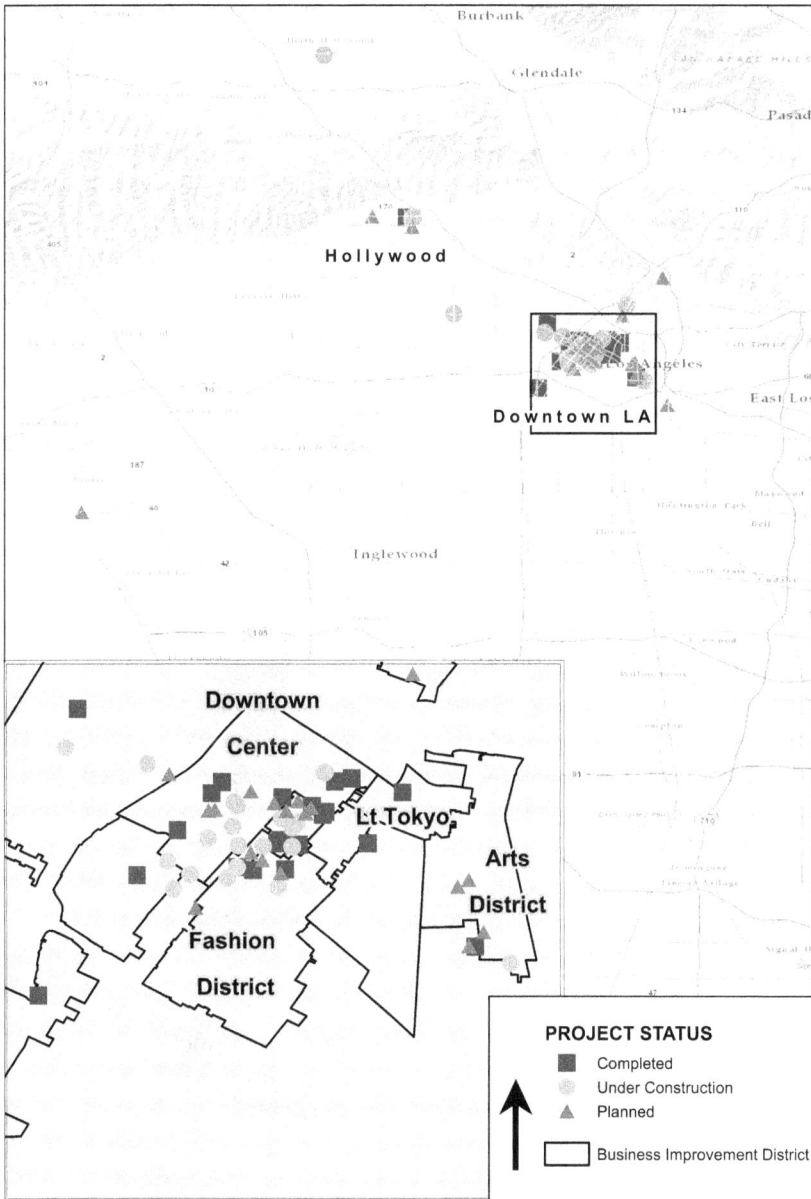

FIGURE 10.1 Adaptive re-use projects in Los Angeles (historical Downtown and Arts
District)

Source: Thomas Sigler, School of Earth and Environmental Sciences (SEES), UQ.

residents to return and live in the downtown are necessary. The application of
adaptive re-use to a building can range significantly from aesthetic modification to
extensive overhaul of buildings, which includes partial reconstruction of the building
while maintaining a façade. Reinterpretation of uses through adaptive re-use is often

assumed by developers to increase or maximize the value of a particular property. Adaptive re-use can achieve both gains in urban sustainability and preservation of cultural heritage in parallel, often not an easy feat to accomplish using other kinds of planning tactics. The success of adaptive re-use in LA was primarily due to the abundance of office buildings available for redevelopment in the historic core of the City.

Innovation process

In this section, we identify the contextual and external factors that have contributed to the emergence and implementation of this innovation. It is important to note that adaptive re-use in LA has been implemented through the ARO and the live/work ordinance; the focus of this chapter is on the former, as it applies to the historical core, which had the greatest potential in terms of adaptive re-use (see Figure 10.1). The first aspect to address is that the planning innovation emerged from a lengthy process involving different groups of stakeholders. Different perspectives on revitalization contributed to the development of the planning innovation, as explained in the following section.

Urban renewal was quite common in American cities during the 1970s. By the early 1970s, urban renewal had already resulted in the clearance of Bunker Hill and new development of skyscrapers was taking place there and in portions of Downtown, south of Bunker Hill, where private demolition of (usually small-scale, but still historic) buildings for parking had already begun (City of Los Angeles Department of City Planning 2005).

In 1976, the landmark California State Historical Building Code (SHBC) was passed, decelerating the demolition that was otherwise normal during this time (City of San Diego Planning Department 2016) and encouraging adaptive re-use practice within the state. The SHBC was unique in the sense that it acknowledged the specific challenges common to historic buildings vis-à-vis modern building codes and incorporated building regulations that were targeted to the rehabilitation and restoration of these buildings (California Office of Historic Preservation 2013). In recognizing these unique construction issues, the state could provide a means to protect its architectural heritage and allowed Los Angeles to retain one of the nation's most intact historic downtown areas. However, the key factor in the absence of demolition was the lack of interest in doing much of anything with these buildings or that part of Downtown at all.

The SHBC can be used by buildings that are locally designated as historic or listed in the National Register of Historic Places. Work undertaken is required to conform to the Secretary of Interior's Standards for the Rehabilitation of Historic Properties, and any new construction undertaken must conform to the State of California's normal code. The SHBC is performance-oriented, as it gives property owners the flexibility to pursue more cost-effective ways to rehabilitate historic buildings.

Preservation of historic buildings provides a variety of economic advantages that often later leads to increased investment in areas where the preservation occurs. One notable example of the successful economic stimulus provided through historic preservation incentives is the Federal Historic Tax Credit (HTC). Passed in 1976, the HTC has cumulatively induced private sector development of US$109 billion in rehabilitation of approximately 39,600 historic buildings (National Trust for Historic Preservation 2014). Additionally, it has generated 2.4 million jobs and spurred the renovation and creation of 450,000 housing units nationally. One prominent project in downtown Los Angeles supported by the HTC is the formerly vacant Far East Building, rehabilitated in 2002-2003. The HTC provided $600,309 in financing towards the project to restore the building's structural integrity and convert 24 single-room-occupancy units into 14 affordable studios and 2 one-bedroom units (National Trust for Historic Preservation 2016).

By the late 1990s, historic preservation was viewed as a prudent way to encourage economic revitalization (Bernstein 2012). In April 2000, a study was carried out by the Los Angeles Conservancy to shift its policy focus to housing creation in the historic core area. A task force was commissioned that was comprised of architects and engineers who were to assess the re-use potential of 273 buildings in the historic core. Later known as the Historic Core Housing Survey, the study identified 50 buildings that were suitable for conversion to housing as articulated in the Broadway Initiative (Young 2009). The significance of the Historic Core Housing Survey is its role as one of the first real catalysts for demonstrating the great untapped potential inherent in Downtown LA's historic buildings.

The success of the adaptive re-use strategy relied on the entrepreneurship of developers (such as pioneer Tom Gilmore from New York). However, our research has clearly identified that the strategy was also made possible through financial incentives and through the involvement of a specific set of stakeholders; the conversion of buildings is a process that has been refined over the years. The innovation process is embedded in LA's political and planning context and is the result of lengthy negotiations between different groups of stakeholders.

In 1993, the Community Redevelopment Agency (CRA/LA), in a public-private partnership for the historic Grand Central Market (bought by a private developer in 1984), worked with the developer to recondition and reposition the building and three former office buildings into new transit-oriented development. The initial concept was substantial rehabilitation and refurbishing of the 1917 market and upgrading the upper floors of three buildings into office space. As it quickly became clear that there was essentially no market for office space in those historic buildings (there was substantial office space available in modern office buildings in adjacent Bunker Hill), the developer proposed converting two of the three buildings above and adjacent to the Market to housing. The conversion proved extremely costly and time consuming, even with the SHBC, due to local code requirements that demanded numerous zoning variances. This experience triggered interest in the ARO, which waived many requirements, such as building setbacks and on-site parking.

The ARO streamlined the building conversion process to attract more developers into downtown renewal projects. The first project carried out under the ARO was undertaken by Gilmore Associates in what is now touted as the highly successful

Old Bank District. The project, which opened in 2001, consists of four formerly abandoned and historic office buildings transformed into over 230 rental apartments with commercial uses on the ground level. These buildings include the San Fernando building (1907), the Hellman Building (1902), the Continental Building (1904), and the Farmers and Merchants Bank (1905). Following its success, several other developers quickly followed suit in the re-use movement in the area and the Old Bank District project would act as the primary precedent and catalyst for further redevelopment that would spur the dramatic revitalization of Downtown. However, we identify through our research that this kind of success was not guaranteed.

Other developers were stymied by conservative financial institutions uncomfortable with financing buildings that did not provide parking at typical Southern California levels. The CRA/LA brokered agreements between residential developers and publicly developed parking facilities to earmark parking for a number of adaptive re-use conversions. It was not until the market demonstrated that historically common levels of parking were not needed or desired by potential residents that there was no longer a need to intervene in that manner.

The Los Angeles Conservancy played a major role in implementing and marketing the ordinance. In 2002 and 2003, the City and the Conservancy organized roundtables to persuade lenders to provide money to developers willing to use the ordinance. This marketing effort benefitted other cities. Although others (Denver, CO and Portland, OR for example) had undertaken adaptive re-use projects before Los Angeles, issues of parking, for example, were of less concern. Later projects in other cities benefitted from the analysis done by CRA/LA and the Conservancy's marketing efforts in demonstrating the viability of adaptive re-use in more urban settings (Lopata 2017).

The Gilmore project also benefitted from public funding starting with the City (via CRA/LA) intervening to prevent an auction of the notable Continental Building for failure of a prior owner to stay current on property taxes, which investment leveraged local, state, and federal funding to make the project feasible. The City committed federal Community Development Block Grant funds; other projects used funds budgeted by the State of California under its Downtown Rebound Program (adopted in 2000), legislation sponsored jointly by Los Angeles, San Francisco, San Diego, and San Jose, CA (California Department of Housing and Community Development 2002). If we refer to the work of Loorbach and Rotmans (2010), the ARO and the different financial incentives described above provided the necessary support for the "go-getters" (the developers) to initiate the transition process. In this process, developers like Tom Gilmore were frontrunners and a project like the Hellman Building's conversion that started in 1998 and completed in 2001 can be considered as a breakthrough project. It was a starting point for a longer-term sustainability vision for Downtown LA.

Trade-offs

With all the benefits that the ARO delivered, the introduction of the ARO did not come without some consequences. In stimulating the revitalization of

Downtown Los Angeles, the ARO has raised concerns related to the gentri-fication of Downtown. While buildings using the ARO were largely vacant office buildings and did not directly displace occupants, low-income residents have still increasingly become displaced and new affordable housing develop-ments have been sparse (Bernstein 2012; Twigge-Molecey 2009). With the notable increase in levels of investment from developers undertaking adaptive re-use projects Downtown, a new demographic has been introduced into the neighbourhood and radically altered many areas into becoming more chic urban spaces. Yung and Chan (2012) identify the challenge associated with generat-ing sustainable development through adaptive reuse while also retaining social inclusiveness and cohesion.

The large upfront cost inherent to adaptive re-use initiatives often causes devel-opers to increase rental fees significantly in an effort to make their money back in a shorter period of time (Smallwood 2012). Out of 9,000 units, only 797 units were produced as affordable housing (Bernstein 2012). Nonetheless, the ARO has proved to be an exceedingly powerful regeneration tool. The City has seen a notable increase of 4,146 residential units in 2010 to 11,733 in 2014, and these numbers continue to rise over time (City of Los Angeles 2015). Gentrification and lack of affordable housing appear as quite common concerns with a variety of successful adaptive re-use projects, and the adaptive-reuse in Downtown LA is no exception. Rapid physical transformation in historic core areas tends to prompt gentrifying features with the often-inevitable increase in real estate val-ues. Given that the primary motive of the ARO was the preservation and re-use of historic buildings, there were not clear strategies implemented to mitigate the negative effects of gentrification and limited affordable housing shown in this LA case study. In addition, at the time, the CRA/LA was still actively engaged in its legally mandated affordable housing production role, focused on new con-struction and renovation and retention of existing housing in the Downtown, rather than on adaptive re-use projects, to meet its housing mandate (Community Redevelopment Agency of the City of Los Angeles 2010). A combination of lack of prioritization in providing affordable housing at the time with pursuit of bottom line financial incentives related to the redevelopment of Downtown LA taking precedence over social components can be seen as primary factors con-tributing to weak social components. A key lesson learned is that when adaptive re-use projects become largely successful in short amounts of time, it is more difficult to introduce policies and incentives after the fact to mitigate these issues. Early introduction of policies and incentives to maintain a certain percentage of affordable housing units in re-use projects may be beneficial as one type of core mitigation strategy.

Conclusion

Los Angeles has been described as having experienced an "urban renaissance" with the profound revitalization of Downtown (Marquardt and Fuller 2012, 155).

The successful urban regeneration of the central city produced in part through the use of the ARO acts as a valuable precedent for many other cities to employ similar adaptive re-use tactics.

The involvement of several stakeholders and use of various tools and policies was key to advancing the goals of historic preservation and revitalization of Downtown. This confirms that innovation for this case study relied mainly on contextual factors such as the development of networks of actors pursuing the adaptive re-use agenda. As we have shown, local financial incentives also helped "go-getters" like Tom Gilmore produce adaptive re-use projects that were commercially viable. The ARO eased many other non-building code constraints such as residential yard and setback requirements (not required of commercial buildings but otherwise needing waivers when commercial buildings were converted to housing), on-site open space (another residential requirement not otherwise addressed in a building conversion), and fairly substantial on-site parking requirements. The last was very significant as parking had to be on the same or adjacent lot (commercial parking can be remote) and interstitial parking lots, though common in Downtown, were not sized or located to meet the parking code.

The enactment of the ARO legislation itself in 1999 took several years of research, education, and outreach by the City and the non-profit Central City Association. The active role of the non-governmental Los Angeles Conservancy in creating the Broadway Initiative to further highlight the untapped potential of Downtown cannot be understated. The application of historic preservation incentives such as HTC and mechanisms such as the SHBC in Downtown demonstrates the importance of incentives as a support tool in urban regeneration. The success of the adaptive re-use strategy relied on the entrepreneurship of developers (such as the pioneer Gilmore), but our research has clearly identified that the strategy was also made possible through financial incentives (at least in the initial stages before market feasibility was revealed) and through the involvement of a specific set of stakeholders that can be considered as a network of actors built around the sustainability transition agenda.

While recognizing the success of adaptive re-use in revitalizing Downtown Los Angeles, LA was a relative latecomer to the adaptive re-use movement. Denver, CO and Portland, OR initiated adaptive re-use in, respectively, Denver's Lower Downtown and Portland's Pearl District, though both initially – as was the case in LA's Arts District – focused on the repositioning of derelict industrial structures (Jackson 2015; Shipley 2012). Indeed, various cities around the world (Berlin, Beijing, Miami, etc.) have taken similar steps to reposition abandoned and obsolete industrial buildings; such repositioning is now a common theme in many urban centres.

More comparable to the Downtown LA experience – certainly one of the largest cities to move into adaptive re-use, demonstrating its potential success in cities of scale – are recent conversion efforts in Chicago, Lower Manhattan, and Detroit, MI. One specific entrepreneurial developer-investor (Dan Gilbert) has acquired – initially at very favourable prices, given that Downtown Detroit had been largely abandoned by developers and investors – and repositioned over a dozen landmark Detroit office buildings, filling them with a variety of companies he controls, starting with Quicken Loans (Detroit 2.0 2016). This created a "critical mass" of occupancy that other developers have followed. A specific contribution of the Los Angeles experience was to demonstrate the potential re-use of office structures (in contrast to industrial buildings) to create new urban infill housing, taking advantage of under-utilized pre-existing infrastructure (streets, sidewalks, water, power, and sewerage systems). Indeed, developers in Downtown Detroit after Gilbert have acquired buildings for residential and hotel adaptive re-use, seeing the potential for occupancy with an increasing work force brought in by Gilbert's efforts.

Common characteristics in all cases are a supply of vacant or largely vacant buildings of historic and architectural merit, cooperative local governments willing to provide regulatory relief, relatively low building acquisition costs, supportive institutions and non-governmental organizations. A recognized housing demand (largely housing in Denver, Portland, and Manhattan, and hotels in Chicago) or, in the case of Detroit, demand created by Dan Gilbert's companies and their work forces, is also a common characteristic. The presence of a young urban population less willing to commute from suburban locations, committed to an urban setting, and largely self-employed or working from home further creates a sizeable demand, especially for urban infill housing. Other central cities with similar characteristics – an available stock of buildings, willing investors and developers, appropriate regulations or a regulatory environment willing to make needed accommodations for adaptive re-use, institutional support, and a potential user population – should be able to replicate the positive results of the locations cited above. North American cities, in particular, are best suited to replicate this planning innovation given the close similarities in their zoning and other land use controls. These cities should work to leverage the appropriate building stock for adaptive re-use, facilitate early stakeholder engagement with a variety of actors, incentivize the entrepreneurship of developers, and explore potential public-private partnerships. One common contextual factor was the latent demand for reconversion in these cities.

Overall, the analysis of the case study highlights the prevalence of the contextual factors in the development of the planning innovation. Even if the adaptive re-use movement in LA benefitted from external sources of funding such as the Federal Historic Tax Credit, most of the financing incentives to support the action of developers emerged at the city scale through newly created networks of the actors built around the adaptive re-use agenda.

References

Austin, R. L., Woodcock, D. G., Steward, W. C. and Forrester, R. A. (1988) *Adaptive Reuse: Issues and case studies in building preservation*. Van Nostrand Reinhold, New York

Avila, E. (2004) *Popular Culture in the Age of White Flight: Fear and fantasy in suburban Los Angeles*. University of California Press, Berkeley and Los Angeles, CA

Banuelos, R. J. (2014) "City of Los Angeles Arts District Form-Based Code", Master's Thesis, California Polytechnic State University

Berman, M. (1982) *All That Is Solid Melts into Air: The experience of modernity*. Verso, London and New York

Bernstein, K. (2012) "A planning ordinance injects new life into historic downtown", in Sloane D. C. (ed.), *Planning Los Angeles*. American Planning Association, Chicago

Bullen, P. A. and Love, P. E. (2010) "The rhetoric of adaptive reuse or reality of demolition: Views from the field", *Cities*, 27, 4, 215–224

California Department of Finance (2013) "Historical Census Populations of Counties and Incorporated Cities in California, 1850-2010", (www.dof.ca.gov/research/demographic/state_census_data_center/historical_census_1850-2010/view.php) accessed 18 May 2016

California Department of Housing and Community Development (2002) "Downtown Rebound Capital Improvement Program 2000", Department of Housing and Community Development, Sacramento, (www.hcd.ca.gov/financial-assistance-downtown-rebound-capital-improvement-program/) accessed 26 May 2016

California Office of Historic Preservation (2013) "Incentives for Historic Preservation in California", Technical Assistance Series # 15, California Office of Historic Preservation

City of Los Angeles (2006) "Adaptive Reuse Program", (https://www.downtownla.com/images/reports/adaptive-rescue-ordinance.pdf) accessed 9 May 2016

City of Los Angeles (2015) "Economics and Demographics. Appendix A: City of Los Angeles Information Statement", (http://cao.lacity.org/misc/AppendixA.pdf) accessed 12 May 2016

City of Los Angeles Department of City Planning (1970) "Concept Los Angeles: The Concept for the Los Angeles General Plan", City of Los Angeles Department of City Planning

City of Los Angeles Department of City Planning (1998) "City Planning and Planners in Los Angeles (1781–1998)", City of Los Angeles Department of City Planning

City of Los Angeles Department of City Planning (2005) "Central City Community Plan", City of Los Angeles Department of City Planning

City of Los Angeles Department of City Planning (2017) "Policy Initiatives", City of Los Angeles Department of City Planning, (http://cityplanning.lacity.org) accessed 10 September 2017

City of San Diego Planning Department (2016) "Find out About the California Historical Building Code", (www.sandiego.gov/planning/programs/historical/faq/code.shtml) accessed 15 May 2016

Community Redevelopment Agency of the City of Los Angeles (1998) "History of Skid Row Series, Paper 1: CRA's Role in the History and Development of Skid Row Los Angeles", Community Redevelopment Agency of the City of Los Angeles

Community Redevelopment Agency of the City of Los Angeles (1982) "Downtown Los Angeles: From the Los Angeles Times", Community Redevelopment Agency of the City of Los Angeles

Community Redevelopment Agency of the City of Los Angeles (2010) "Building a World Class City", Community Redevelopment Agency of the City of Los Angeles

Darchen, S. (2016) "Regeneration and networks in the Arts District: Rethinking governance models in the production of urbanity", *Urban Studies*, 54, 15, 3615–3635.

Demographia (2016) *World Urban Areas*, 12th Annual Edition, (http://www.demographia.com/db-worldua.pdf) accessed 12 May 2016

Detroit 2.0 (2016) "Crain's Detroit Business", (www.crainsdetroit.com/taxid/19006063/detroit-2-0) accessed 22 June 2016

Downtown Center Business Improvement District (2014) "Reaching Higher: Downtown Center Business Improvement District 2014 Annual Report" (https://www.downtownla.com/images/about/DCBID_2014_Annual_Report_updated.pdf) accessed 12 May 2016

Gordon, P., Richardson, H. W. and Wong, H. L. (1986) "The distribution of population and employment in a polycentric city: the case of Los Angeles", *Environment Planning A*, 18, 2, 161–173

Jackson, M. (2015) "Before and After: Ten Examples of Adaptive Re-Use in Denver, CO", Confluence Denver (www.confluence-denver.com/features/adaptive_reuse_101415. aspx) accessed 12 May 2016

Joh, K., Chakrabarti, S., Boarnet, M., and Woo, A. (2015) "The walking renaissance: A longitudinal analysis of walking travel in the Greater Los Angeles Area, USA", *Sustainability*, 7, 7, 8985–9011

Langston, C. A. (2008) "The Sustainability Implications of Buildings Reuse", CRIOCM International Research Symposium on Advancement of Construction Management and Real Estate, Beijing, China

Loorbach, D. and Rotmans, J. (2010) "The practice of transition management: Examples and lessons from four distinct cases", *Futures*, 42, 237–246

Lopata, J. R. (2017) "The Rise of Los Angeles 3.0", Senior Honors Thesis, Department of History, Stanford University

Los Angeles Conservancy (1999) "Broadway Initiative Action Plan", Los Angeles Conservancy

Marks, M. A. (2004) "Shifting ground: The rise and fall of the Los Angeles Community Redevelopment Agency", *Southern California Quarterly*, 86, 3, 241–290

Marquardt, N. and Fuller, H. (2012) "Spillover of the private city: BIDs as a pivot of social control in downtown Los Angeles", *European Urban and Regional Studies*, 19, 2, 153–166

National Trust for Historic Preservation (2014) "The Federal Historic Tax Credit: Transforming Communities", (www.preservationnation.org/take-action/advocacy-center/policy-resources/Catalytic-Study-Final-Version-June-2014.pdf) accessed 19 May 2016

National Trust for Historic Preservation (2016) "Far East Building", (www.preservation nation.org/information-center/economics-of-revitalization/rehabilitation-tax-credits/jobs/far-east-building.html#.Vz4JfZQYOM8) accessed 19 May 2016

Shipley, K. (2012) "Adaptive Re-Use in the Pearl District, Portland, OR", Prezi Presentation, October 13.

Smallwood, C. (2012) "The role of adaptive (re)use", PMI Educational Foundation for Student Paper Competition, 1–7 (https://www.pmi.org/learning/library/role-adaptive-reuse-6010)

Twigge-Molecey, A. (2009) "Is Gentrification Taking Place in the neighbourhoods surrounding the MUHC?", *CURA Making Megaprojects Work for Communities*, 1–80 (https://secureweb.mcgill.ca/urbanplanning/files/urbanplanning/RR09-02E-twigge.pdf)

United States Census Bureau (2016) *"Population and Housing Unit Estimates"*, (www.census. gov/popest/) accessed 15 May 2016

Whittemore, A. H. (2012) "Zoning Los Angeles: a brief history of four regimes", *Planning Perspectives*, 27, 3, 393–415

Young, M. A. (2009) "Note: Adapting to adaptive reuse: Comments and concerns about the impacts of a growing phenomenon", *Southern California Interdisciplinary Law Journal*, 18, 703–706

Yung, E. H. and Chan, E. H. (2012) "Implementation challenges to the adaptive reuse of heritage buildings: Towards the goals of sustainable, low carbon cities", *Habitat International*, 36, 3, 352–361

11

BHUBANESWAR, INDIA

Smart City Plan and economic sustainability

Tathagata Chatterji and Saugata Maitra

Introduction

This chapter analyses the key innovative features of the Smart City Plan (2015) of Bhubaneswar – capital of the eastern Indian state of Odisha. The plan is closely associated with the economic restructuring of the city and aims at transforming a sleepy administrative town into a vibrant regional hub of business services and tertiary education.

The Smart City Mission was launched by India's national government in 2014–15, with the ambitious aim to lift up economic potential of 100 cities across the country through techno-managerial interventions. Bhubaneswar's Smart City planning proposal ranked first in the nationwide competition to fund cities under the mission – due to its innovative approach towards participatory planning and extensive community engagement mechanism. Subsequently, the plan also received the Pierre L'Enfant International Planning Excellence Award 2017 from the American Planning Association, for articulating a citizen-driven vision for the future through use of modern technology (APA 2017).

Bhubaneswar's planning proposal redefines the concept of 'smart cities' and outlines a citizen-driven vision for the future by using technology to help residents gain better access to city services, and improve the overall quality of life (BSCL 2015). The hallmark of the plan is the customized and nuanced approach towards citizen engagement. Targeted application of face-to-face and social media platforms enabled the plan to reach out to 32 per cent of the city's 0.97 million inhabitants and garner support behind the plan from a wide cross-section of the population, cutting across socio-economic segments, gender and age groups. Consultation with a diverse range of stakeholders had produced a plan based on broad consensus, which takes a balanced, moderated approach towards urban development.

The prime focus of the Smart City Plan (2015) is to achieve systemic efficiency through governance process reengineering and improve coordination between the multitude of public and private agencies involved in urban management. The plan takes forward the Odisha state government's economic strategy to position Bhubaneswar as the leading knowledge economy hub amongst the secondary cities of India, by targeting interventions in three core areas: institutions and regulations; land and infrastructure; and skill development and training.

First, the plan attempts to streamline inter-agency coordination through integrated management information systems, data-sharing and transaction platforms. A new institutional architecture has also been put in place for single-point coordination of all Smart City related projects. Second, the Smart City Plan brings in a new regime of operational efficiency in the delivery of civic services, including reduced pilferage, transmission and distribution losses, through service-level benchmarking and real-time monitoring. Third, the plan complements the state government's efforts in attracting knowledge economy investments, by focusing on employability and entrepreneurship of the urban youth through skill development centres and micro-business incubators.

The Smart City Plan lays the road map for Bhubaneswar's economic development trajectory to be inclusive, resource efficient and technology enabled. The next section provides a background of the city, its evolution and its planning history. The third section discusses sustainability challenges faced by the city in recent years. The fourth section then analyses the Smart City Plan, including its innovative approaches and trade-offs associated with the innovation process. A final section wraps up the discussion by highlighting the lessons learnt and in what way a similar initiative could be replicated in other cities.

The city: Bhubaneswar

Bhubaneswar, the capital of Odisha, is an emerging information technology (IT) and education hub of eastern India. The core part of the city under Bhubaneswar Municipal Corporation (BMC) has a population of 0.97 million and covers an area of 161 sq. km (BDA-BMC 2015). Bhubaneswar Development Authority, an agency directly under the state government, is responsible for urban planning activities for the entire city-region (Bhubaneswar Development Plan (BDP) Area), which includes the municipal area and its peri-urban fringe, covering about 1100 sq. km and having a population of 1.3 million (BDA-BMC 2015). Figure 11.1 shows the municipal core and the city-region.

The modern city of Bhubaneswar was developed in 1948–49, to move administrative functions away from the congested, commercial town of Cuttack, situated on the opposite bank of the river Daya. The new city was located beside a historic settlement with several architecturally significant temples dating back to the eleventh century. It was designed by German architect Otto Koenigsberger as a spread out, low-density, grid-iron layout, based on the neighbourhood concept and it acted as a precursor to Chandigarh (Kalia 1995).

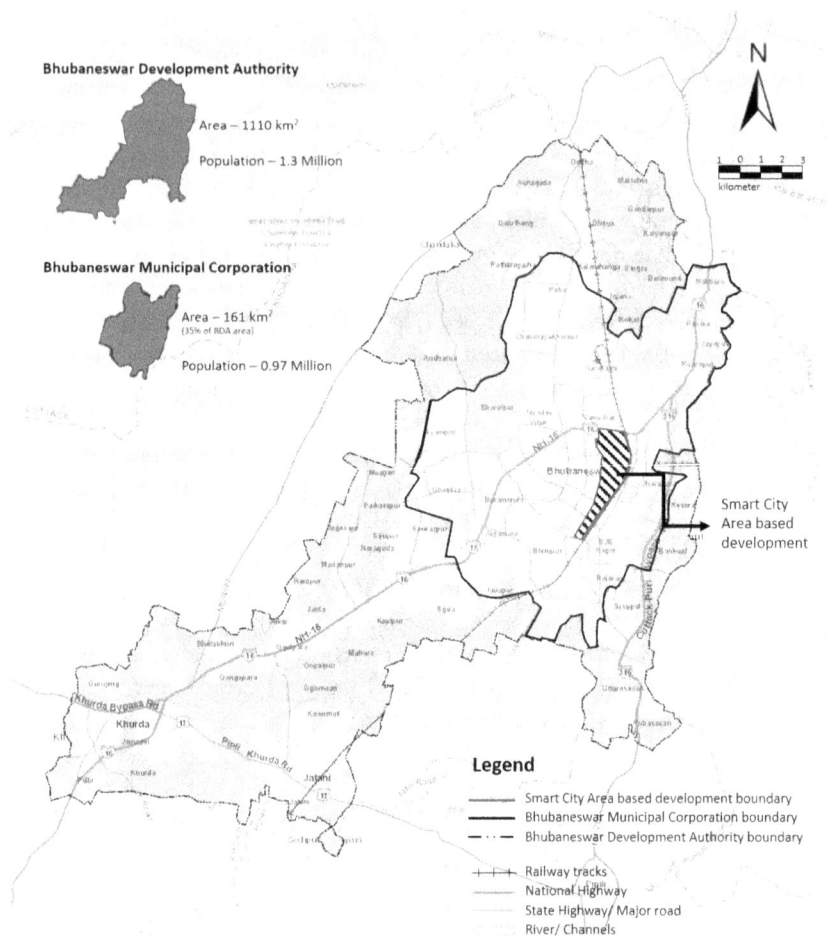

FIGURE 11.1 Bhubaneswar showing BDP and BMC areas

Source: Modified by authors based on data sourced from BMC-BDA Smart City Plan (2015)

Bhubaneswar's population increased from 16,500 to 837,700 between 1951 and 2011 (Government of India 2011). During each of the first three decades (1951–61, 1961–71 and 1971–81), the city witnessed a growth rate over 100 per cent as various government departments began to consolidate their operations in the newly built state capital; significant expansion in state apparatus also took place in the era of public sector-led development. Private investments were mainly in the tourism sector. Following the improvement of the transportation network, Bhubaneswar emerged as the gateway to the cultural tourism circuit connecting the temple towns of Puri and Konark – a UNESCO World Heritage site.

From the early 2000s, the economy of Bhubaneswar started to diversify as the Odisha state government stepped up efforts to attract private investments by improving infrastructure and policy reforms. Odisha has rich natural resources (coal, iron, bauxite, manganese etc.), yet long remained industrially backward and economically poor. However, over the past two decades the state enjoyed political stability and made steady progress in terms of economic and social development indicators. Amongst 32 states of India, Odisha now ranks 7th in the Ease of Doing Business Index (Centre for Civil Society 2015), 14th in cumulative FDI (foreign direct investment) inflow between 2000 and 2016, 26th in per-capita state gross domestic product (GDP) and 22nd in the Human Development Index (Ministry of Finance 2017; MoSPI 2017). Odisha still continues to be an overwhelmingly rural state, with only a 17 per cent urbanization level – which is well below the national average of 31 per cent (Government of India 2011).[1] Nevertheless, the state capital Bhubaneswar has become a prominent business destination amongst the second tier cities in India.

Bhubaneswar had been particularly successful in attracting investments in the IT-enabled back offices and tertiary educational institutions. The economic geography of the Indian IT sector started to change from the early 2000s onwards, as big IT companies such as Infosys, Wipro, TCS etc. started to relocate their back office and lower end functions to second tier cities, to offset rising land and labour costs in established IT clusters (e.g. Bangalore, Pune etc.). Moreover, state governments stepped up their efforts to attract IT sector investments, by offering subsidized land and various fiscal incentives, due to increasing domestic economic competition (Chatterji 2017).

Bhubaneswar's locational proximity to the Kolkata metropolitan region, with a large educated population base, also played a key role in Bhubaneswar quickly climbing up to ninth position nationally in terms of IT services exports (NASSCOM 2016). Besides IT services, Odisha also received substantial investments in mining and manufacturing sectors. Companies engaged in mineral resource extraction, power generation, steel production, aluminium processing and various other industrial activities elsewhere in the state, opened their regional offices in the state capital.

The education segment has also expanded substantially, as several reputed institutes (e.g. Indian Institute of Technology, Indian Institute of Information Technology, All India Institute of Medical Science, Xavier Institute of Management etc.) opened campuses in the city. There are now over 100,000 students in the city, from across the country. People in the 15–24 years age group account for 20.07 per cent of the city population (Government of India 2011). Multiplier effects of the knowledge services production economy have contributed to further downstream investments in retail, housing, specialty medical facilities and hotels.

The Smart City project is further expanding the agglomeration advantage of Bhubaneswar's knowledge economy by facilitating new public sector investments in hardware and software for e-governance and urban management information systems. Companies such as Honeywell Technologies, who had no presence in the city before, have now opened offices, as they have been entrusted with the contract for managing the Integrated Command and Control Centre under the Smart

Cities project. Growth in technology services is expected to further boost the tertiary education sector to meet the demand for a skilled workforce. Multiplier impacts of growth in the technology sector are also expected to boost corporate real estate sectors (e.g. housing, hospitality, commercial retail etc.) to meet the consumption needs of the workforce.

Urban sustainability challenge: Environmental degradation and socio-economic polarization

Bhubaneswar's strong physical planning legacy of the 1950s took a back seat in the subsequent decades, as the political leadership of the overwhelmingly rural state did not pay much attention to urban issues. Moreover, being a planned and relatively new city, urban blight and hyper congestion were not so apparent (in comparison to other Indian cities) during the first decade of Bhubaneswar's economic reorientation from a government services oriented administrative city to a private enterprise dominated business services hub. However, environmental degradation and socio-economic polarization started to become starker from 2010 onwards, as the city stepped into the second decade of its economic transformation.

First, the city has started to expand outwards along the major highway networks, in the form of disjointed and incoherent ribbon development. IT business parks, big educational campuses and special economic zones are becoming self-contained enclaves at the outskirts of the city, and often outside the municipal limits. Development of such high-end production spaces is in turn spurring demand for consumption spaces, in the form of gated apartment complexes and shopping malls. Speculative real estate investments through monetization of peri-urban agricultural land in a haphazard manner have started to impact the fringe areas of the city. Unplanned development in a piecemeal manner, without provision of trunk infrastructure (e.g. drainage, sewerage and water supply) and indiscriminate construction activities through filling up of paddy fields and rural water bodies are leading to flood hazards, depletion of ground water and loss of green cover and fertile agricultural land.

Second, the rise in middle-class consumerism is putting pressure on the transportation infrastructure of the city and its outer fringe. Disjointed and inadequate public transport services in the newly developing peri-urban areas are encouraging automobile usage. Within the core part of the city, the areas adjoining the secretariat and administrative precincts are seeing high demand for retail and commercial facilities. The low-rise low-density urban form of Bhubaneswar is not conducive to efficient mass transit and vehicular traffic is growing at an unprecedented scale. Roads of Bhubaneswar, being a planned city, are wider than most Indian cities. However, the exponential rise in the number of cars and two wheelers is leading to traffic congestion. Traffic analysis reporting shows bus services, the principal mode of public transport in the city, account for only an 11 per cent trip share, while the overall contribution of public transport is only 16 per cent. Two-wheeler vehicles, which account for 33 per cent of the trip share, are the dominant mode of transport. In total, personal vehicles account for a 40 per cent trip share. Consequently, air quality has started to worsen.

Third, urban economic divides are widening, as disconnection between the GDP growth rate and employment generation is becoming more and more apparent. Like most other Indian big cities, Bhubaneswar is also witnessing that a knowledge economy centric growth strategy leads to a rapid rise in export earnings. But such sectors have a low spill-over effect (Chatterji 2013). IT services and the tertiary education sector are creating employment opportunities for the educated middle class, but not for the urban poor and the rural migrants. As a result, Bhubaneswar is seeing a sharp rise in informal sector trading activities. About 55 per cent of the workforce is engaged in the informal economy and the unemployment rate of 4.27 per cent is marginally above the national average (BSCL 2015; Ministry of Finance 2017). Thus, employment generation for the city's burgeoning youth population is emerging as a core challenge for the state government and is likely to impact future planning policies.

Planning innovation

Smart city initiatives in India have started to draw academic attention – mainly from the angle of social sustainability and inclusivity. Datta (2015) argues that digital technology enabled cities as a new form of utopia, where urbanization is being seen primarily as a business model, rather than as a tool for social inclusion. Focusing on the development process of Dholera, a greenfield port city in Gujarat, from a political-economic perspective, she claims that such utopian cities are being fuelled by a combination forces of 'technocratic nationalism' of the aspirational middle class and 'entrepreneurial urbanism' of the provincial political elites (Datta 2015). Looking from the angle of social inclusion, Roy (2016) critiques the selection process of the cities under the Smart Cities Mission. Performance-based selection criteria had led to selection of cities which are relatively better-off in terms of economic performance and governance competence, rather than improving governance capacities of backward cities (Roy 2016). Praharaj et al. (2018) analyse the urban governance dynamics in Indian cities which have started implementing the Smart Cities Mission, and highlight the need for structural reforms in the governance arrangements to make it more participatory.

While the above arguments and apprehensions are valid, this chapter takes a different approach to discuss how, operating within the framework of the Smart City Mission of the national government, Bhubaneswar increased opportunities for civic participation through its innovative planning process,.

Innovation process

A new opportunity to address Bhubaneswar's urban planning issues and its infrastructure shortfalls came after the national government's Ministry of Housing and Urban Affairs (MHUA) launched the Smart Cities Mission in 2015 and announced a nationwide competition to select cities for funding under the scheme (Ministry of Urban Development 2017). Bhubaneswar ranked first in this nationwide Smart City proposal competition due to its innovative participatory planning – a citizen-driven vision for the future by harnessing technology to help people improve the quality of life. It was a joint initiative involving municipal government, state government and international consultants – Jones Lang LaSalle (JLL) and IBI Group. The Smart City Plan also illustrates how a locally conceptualized innovative planning approach can be dovetailed within the structured framework of a centrally sponsored scheme.

The Smart City Mission is a financial assistance package as well as a policy tool kit. The overarching aim of the mission is to rejuvenate 100 Indian cities through structured, techno-managerial interventions over five years (2015–16 to 2019–20). It has a budget of US$15 billion (Ministry of Urban Development 2017). Cities were selected through an elaborate point-based two-stage competition. The first stage was intra-state and then the shortlisted cities competed nationally. Cities qualifying through this process are entitled to a grant of 50 per cent of the project cost, with a maximum limit of INR500 cr ($79 million) from the MHUA. An equivalent amount is to be provided by the state and municipal governments. The MHUA grant included a budget of INR2 cr ($316,000) to each city to develop its project proposals by engaging reputable consulting firms. Each city's proposal was required to comprise the following two components:

- Pan City – this component aims to improve delivery of a specific infrastructure network across the city through big data-intensive smart technology.
- Area Based Development – this component provided opportunities for retrofitting of an existing built environment spread over 500 acres through smart technology; redevelopment of an existing neighbourhood of 50 acres to promote high density mixed use development; and greenfield development of a minimum 250 acres through innovative and sustainable planning (Ministry of Urban Development 2017).

For efficient and fast track project implementation, all cities selected under the Smart Cities Mission are required to set up a Special Purpose Vehicle (SPV). The SPVs are set up as registered public sector companies with scope for private sector participation. State government agencies and municipal governments are together required to hold 60 per cent of the equity, while private sector equity participation is allowed up to 40 per cent. In keeping with this requirement, Bhubaneswar Smart City Ltd (BSCL) has been formed. It has now been entrusted as a single point of contact to handle the Smart City project development, procurement, infrastructure, utilities management and services.

The Smart City Plan carries forward the aspirational goal of the Odisha state government to improve Bhubaneswar's economic competitiveness and pitch the city as the leading 'Regional Economic Centre' amongst India's Tier-2 cities. The plan augments these economic objectives by encouraging knowledge-based enterprises, start-ups and sustainable tourism. The plan promotes a more 'Liveable City' by providing a diverse range of housing, educational and recreational oppor-tunities and enhancing heritage conservation; providing safe and accessible public spaces to become a 'Child Friendly City'; and promoting an 'Eco City'[2] and 'Transit Oriented Development'.

Under the Pan City component, the Bhubaneswar Smart City Plan proposes a state-of-the-art Intelligent City Operations and Management Centre (ICOMC) that integrates multiple city systems, including traffic management, parking, bus/transit operations, smart utilities and emergency response. Apart from facilitating regular urban management functions through a common platform, ICOMC is also expected to play a crucial role in disaster management.

Other projects under the Pan City component include an intelligent traffic signalling and flow monitoring system; an integrated parking management system, linking public and privately-owned parking spaces in commercial and business districts; and an intelligent bus tracking and fleet management system. To facilitate digital transactions, the city is launching a Smart Card having features of a cash card, but loaded with features to access various public services (e.g. utility bill payment), ticketing in city buses and commuter trains, car parking charges, recharge of mobile phones, amusement and recreational places, cinemas, restaurants and even at some select stores.

The strategy behind the Area Based Development component is to densify the inner core and discourage unplanned expansion and ribbon development. Moreover, by encouraging retail, food court, entertainment and cultural facilities the city is attempting to create spaces attractive to its burgeoning youth population and erase the perception that Bhubaneswar is somewhat 'dull' and 'staid'.

The extensive place-making effort calls for redevelopment and retrofitting of 985 acres around the centrally located transit station into a vibrant 24/7 destination called the Bhubaneswar Town Centre District. The land use plan of the Town Centre District has earmarked 31 per cent of the land for residential usage, 15 per cent for mixed use and 12 per cent for commercial activities, to encourage round-the-clock activities, and avoid the place becoming dead after office hours (see Figure 11.2 for the detailed land use structure).

Moreover, the Railway Station Multi-modal Hub is being designed to have several attractive public spaces such as a theatre plaza, nature plaza, art plaza, town plaza, etc. connecting art galleries, food courts, shopping malls and retail (700,000 sq. ft) and office buildings (110,000 sq. ft), serviced apartments (530 in total), and hotels and convention centres (420,000 sq. ft). The central commercial street, Janpath, with a right of way of 60 m and 5 km length, is being redeveloped, with designated pedestrian walkways, cycle tracks, automobile lanes and bus lanes. High-rise, high-density developments have been planned along the transit corridor, to generate sufficient commercial uses.

The mixed use character of the development, as well as the high proportion of retail and recreation oriented land use, is expected to make the heart of the city livelier and bring in greater economic vibrancy. Following redevelopment and

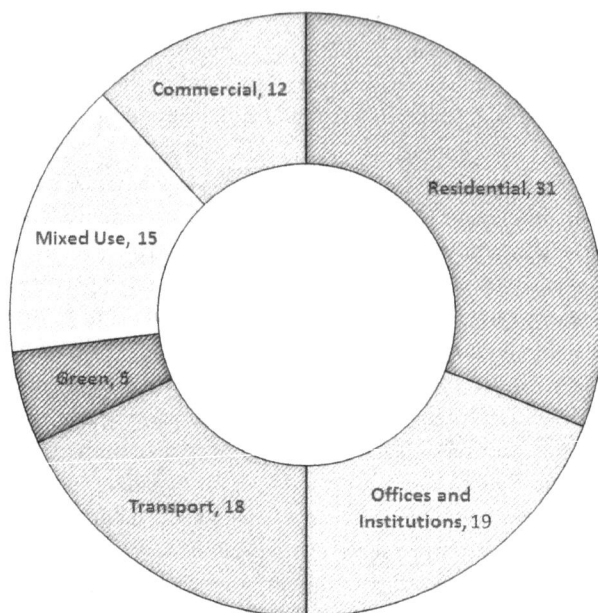

FIGURE 11.2 Town centre area land use (per cent of total for each use)

Source: Prepared by authors based on data sourced from BMC-BDA Smart City Plan (2015)

densification, the Town Centre District is expected to generate additional employment opportunities for about 31,000 people (BDA-BMC 2015).

A particularly noteworthy feature of Bhubaneswar's Smart City Plan was the extensive public consultation process. Based on the 'spectrum of public participation' framework of the International Association of Public Participation (IPA2), it encompassed all five steps: information, consultation, involvement, collaboration and empowerment (BDA-BMC 2015). The communication strategy involved an elaborate audience segmentation exercise involving extensive application of digital media as well as conventional channels. Table 11.1 illustrates in detail the objectives and tools employed in each of the steps.

The multi-pronged community outreach initiative was successful in engaging 32 per cent of the city's population and creating a 'buzz' about the Smart City Plan. The engagement drive 'Citizen Connect Initiative' was a three-month project and obtained inputs from city's residents through discussion forums and face-to-face meetings with slum dwellers and street vendors, online polls, social media outreach, volunteer programs, ideas papers and design competitions. Youth engagement was achieved through street plays, flash mobs and informational kiosks. Smart City Learning Labs, led by moderators, engaged with school and college students. Public personalities joined as campaign ambassadors. The stakeholder interfaces in the engagement process involved discussions with elected representatives, institutional stakeholders, government officials, media,

TABLE 11.1 Citizen engagement strategy

INFORM

Objective	Provide citizens and stakeholders with objective information
Tools	Print media – advertisements, street hoardings
	Electronic media – SMS, mass mailer, radio /TV talks, Facebook, website, advertisements
	Launch program – Citizen Connect

CONSULT

Objective	Obtain feedback from citizens on main challenges and discuss proposed solutions
Tools	Conferences – seminars/techno fairs/stakeholder meetings
	Whatsapp group – focused groups, expert inputs

INVOLVE

Objective	Conduct participatory planning exercises to co-create solutions
Tools	Contests – vision statement, logo design, photography
	Discussions – discussion forums, idea box, community mapping

COLLABORATE

Objective	Partner with individuals, formal and informal groups, to generate ownership in implementation
Tools	Smart City Labs – institutions, schools, colleges, public organizations, slum communities
	Champions – volunteer program, mayor's badge, Smart Lab organization and execution teams

EMPOWER

Objective	To place final decision making in the public hands through citizen juries
Tools	Local democracy – ward level community meetings
	Polling – citywide solutions, area based development, best practices for transport and waste disposal
	Partnership – educational institutions, non-governmental organizations

Source: Prepared by authors based on data available from Smart City Proposal (BDA-BMC 2015)

non-governmental organization (NGOs), slum dwellers, women, youth and children, senior citizens and the physically challenged. The projects suggested for inclusion under the Smart Cities proposal were finalized based on this citizen engagement process. The detailed break-up of people reached through various channels is indicated in Table 11.2.

Overall, Bhubaneswar's Smart City proposal demonstrates a unique and innovative approach towards participatory planning, which sets it apart from plans submitted by other Indian cities. Discussions in the other cities were restricted to few seminars involving a small set of non-state actors. The larger public were only 'informed' through newspapers and other news channels, whereas Bhubaneswar

TABLE 11.2 Citizen's Connect Initiative

	Mode of contact	Number of participants
1	Community meetings/seminars/workshops	
	Number of attendees	310,000
	Number of votes	190,000
	Number of suggestions	160,000
2	Project website	
	Number of visits	170,000
	Number of participants in online polling	29,184
	Number of likes	47,238
3	Facebook page	
	Number of views	3,400,000
	Number of likes	86,851
	Number of comments	1,630
4	SMS	
	Number of messages sent	2,570,000
5	Radio/TV	
	Number of comments	22,674

Source: Prepared by authors based on information available from Bhubaneswar Smart City Proposal (BDA-BMC 2015)

managed to 'involve' one-third of its population in the planning process by creatively leveraging various forms of new and conventional communication channels. The scale and depth of public engagement are particularly noteworthy and unprecedented. As the projects launched under the Smart Cities Mission were generally well received by the different stakeholders, the projects are unlikely to encounter major controversies or confrontational opposition at the time of implementation.

Encouraged by the positive public response during the proposal phase of the Smart Cities project, civic authorities have now sought to further deepen and institutionalize the scope for citizen engagement in everyday urban management. For example, the BSCL has now launched a 'My City My Pride' (MCMP) program to empower citizens to bring about transformation, mobilize the public and connect with the government on a regular basis. MCMP with its mobile app and website allows citizens to view the status of their complaints, learn about other issues in their locality and get attention from the authorities. It is a medium to connect the citizens directly with BMC on issues like health, sanitation, waste management and the sewerage system of their respective areas (BSCL 2017).

Collectively, these projects under the Smart Cities Mission provide the foundation for a more inclusive, resource-efficient and technology-enabled city. In addition, a new form of civic pride has emerged around the Smart City project. The first phase of the project is expected to be completed by the end of 2018.

Trade-offs

The Smart City Plan's Pan City component focuses on improving infrastructure delivery through service level benchmarking and better institutional coordination. An integrated command and control centre in public private partnership (PPP) mode is the lynchpin of the plan. Since the major focus is to improve urban management through internal work process restructuring enabled by software technology, possibilities of adverse environmental fallouts are minimal.

The urban renewal component is restricted to the commercial spine of the city and the railway station area and construction activities are in phases to minimize disruption of on-going local commercial activities. A crucial challenge faced here is the relocation of the informal activities – especially street vendors. The railway station and Janpath road, the main commercial hub of the city taken-up for redevelopment, also happen to be the most lucrative vending zone. According to the estimates of Bhubaneswar Municipal Corporation, there are 1,183 street vendors operating in this stretch (*The Telegraph* 2017).

Urban renewal has become a controversial issue in India, due to livelihood vulnerabilities of the urban poor. 'World class' city-building aspirations of the local political elites and 'clean/green city' drives by environmentalists in several other cities (e.g. Delhi 2008–09, Kolkata 1998–99) had often come about through forcible demolition of slums and eviction of street vendors (Bandopadhaya 2009). This in turn caused much tension – due to agitations, protest movements, and lengthy legal actions.

Bhubaneswar, however, had been following a more conciliatory and participatory approach. 'Negotiating change through minimum friction' and 'Progress through partnership' became the guiding principles of the Bhubaneswar Municipal Corporation from 2006. The municipality took the lead in bringing together various state and non-state stakeholders under a consultative forum umbrella called City Management Group (CMG), and co-opted the vendor unions into it. The CMG through discussions designated vending and non-vending zones across the city; developed special vender markets; erected semi-permanent stalls for vendors along the sidewalks to free up pedestrian movement; and provided licenses to rehabilitated vendors. It is important to note here that India's national parliament enacted the Street Vendors Act in 2014. This act prescribes that street vendors should be included in municipal ward committees, and these committee should be empowered to designate specific areas as vending or non-vending zones. Thus, Bhubaneswar's participatory approach towards relocation of street vendors precedes the national legislation. To facilitate comprehensive urban renewal of a major segment of the commercial core under the Smart Cities project, Bhubaneswar is continuing with the same participatory approach. To minimize disruption of on-going formal and informal economic activities in this busy area, the stakeholders were made part of the consultation loop; the projects are being rolled out in phases; and street vendors are being temporarily relocated to nearby areas.

As per the Smart City Plan, the city administration will develop dedicated and permanent street vending zones in three locations – the Master Canteen square adjoining the Railway Station Multi-modal Hub, Satya Nagar Institutional Area and the Lake District – all within the core Smart City area. Vending zones would be grouped in specific categories, such as a fruit vending zone, handicraft vending zone, mixed vending zone and more. Besides organizing street vending within the central business district (CBD) core, the civic body has decided to allocate uniforms for street vendors across the city and a design competition was organized for this purpose (*The Telegraph* 2017).

While several other cities in India have made provisions for designated vending zones for street hawkers under the Area Based Development component of their Smart Cities proposal, Bhubaneswar's approach towards the issue is the most comprehensive. It is also to a certain extent unique, as it attempts to address the bigger issue of skill building and employability of the urban poor (Ministry of Urban Development 2017). Several skills development centres and micro-business incubators are being created to help residents become better prepared to obtain jobs or open their own businesses. The plan also proposes to empower marginalized sections of society, including slum dwellers, by awarding them property rights. Slum redevelopments would focus on improving basic services and providing opportunities to improve health and nutrition.

Processes of urban economic restructuring in the neoliberal era frequently reconfigure the space relations in cities. Consequently, social divides widen and the environmental conditions worsen – especially in fast-growing Asian cities. Viewed from this perspective, the Smart City Plan is a timely intervention, which shows a technology-enabled roadmap to steer the process of Bhubaneswar's economic transformation in a sustainable manner. The plan attempts to mitigate the adverse environmental fallouts of unplanned automobile-led urbanization, through land policies promoting transit-oriented development, before the situation gets out of control. But more importantly, the economic vision of the city is socially inclusive. Bhubaneswar's aspiration to attract knowledge economy investments and tech-savvy workforce has not overshadowed the stark reality – the need to address livelihood concerns of the urban poor.

By concurrently addressing economic, social and environmental objectives, Bhubaneswar's Smart City Plan struck a balance between competing claims, without any significant adverse trade-off – at least apparently so. Fundamental democratic underpinning of the planning process steered the plan towards a consensual and non-confrontational approach. And here lies the main message: the innovation in the Smart City Plan is not so much in the realm of fancy software or hardware – but rather, creative application of modern communication technology to reach out to diverse sections of the city's population within a short span of time.

Conclusion

This chapter has discussed innovative planning approaches of Bhubaneswar, a medium-sized Indian city. Bhubaneswar's Smart City Plan is closely associated with the city's economic restructuring, from that of a sleepy administrative town to a vibrant regional business services and education hub. The plan redefines considerably the concept of a smart city and outlines a citizen-driven vision for the future through a creative public communication strategy. Our research highlights three major take-away points, which could be applied in other developing countries with suitable customization, based on local contextual factors.

First, the most important hallmark of the plan is the innovative approach towards participation through effective application of online and offline communication tools. A targeted approach of face-to-face communication and social media platforms enabled the plan to reach out to a third of the city's population, comprising diverse socio-economic segments, gender and age groups. The scale and extent of the public outreach was unprecedented in India. The inclusionary process avoided the urban divide and social polarization, which often mark economic restructuring of developing country cities. Bhubaneswar's innovative application of online and offline media platforms calibrated for community engagement shows how modern communication technology can be harnessed in planning communication, and can reach out to diverse sections of the society, including the marginalized. This in turn reduces the scope of hindrances during the project implementation phase.

Second, Bhubaneswar shows a sustainable approach towards urban economic development. Bhubaneswar's Smart City Plan aims to enhance the city's position as the leading knowledge economy hub amongst the secondary cities of India, and attract new investments and aspirational young professionals by improving its quality of life. But its approach towards a technology-driven future is more inclusionary and shows greater focus on the livelihood needs of the urban poor compared to most Indian cities. The city is making strong efforts to improve the economic conditions of the urban poor, through skill building programs. Moreover, urban renewal to meet the needs of the new economy, which often leads to the eviction of the poor in India and other Asian cities, has been dealt with in a more humane manner here, through plans for in situ rehabilitation and temporary relocation.

Third, Bhubaneswar's planning process demonstrates how a locally conceptualized innovative approach could be dovetailed within the structured framework of a centrally sponsored scheme. The impetus for the new plan came from the national government, which also provided 50 per cent of the project funding and the seed capital for planning. The planning process however reflects not only the state government's economic objective, but also its socio-political ideology, in the form of a participatory and inclusionary approach. The planning exercise was carried out by a consortium involving the state government, the municipal government and international consultants. A new institutional architecture has been put in place, in the form of a Special Purpose Vehicle (with scope for private sector equity participation at a later stage) to streamline coordination between various governmental and private agencies responsible for the delivery of urban services.

To sum up, the Smart City Plan shows a technology-enabled participatory roadmap for sustainable economic restructuring of a medium-sized Asian city. This had been made possible due to favourable alignment of institutional, political and economic factors. Strong institutional arrangements, plus inter-governmental networks and relationships, facilitated cooperation between national, state and municipal governments on the one hand and state and non-state actors on the other hand. Moreover, the social-democratic political orientation of the regime in power at the state and city level made it conscious of the needs of the urban poor and thus did not push neoliberal economic approaches too hard. Finally, prudent economic management allowed the city to access various public resources to implement a wide range of developmental actions.

Notes

1 Census classification of urbanization in India has to meet three criteria: population, occupancy and density. Settlements with a minimum population of 5,000, with 75 per cent of the male working population engaged in non-agricultural activities and population density of 400 persons per sq. km are defined as urban by the Census (Government of India 2011).
2 The 'Eco City' component of Bhubaneswar is part a IFC-European Union supported multi-year climate change focused program. It is now aligned with Government of India's Smart Cities initiative as it is structured around designated Smart Cities: Bhubaneswar, Bengaluru, Chennai, Mumbai and Pune. The long-term objective is to help India meet its nationally determined contributions (NDCs) by reducing greenhouse gas emissions and mobilizing private sector finance through a combination of established and innovative interventions designed for the Indian market (Smart Cities Council 2018).

References

APA (2017) *Pierre L'Enfant International Planning Award*, American Planning Association <https://www.planning.org/awards/2017/bhubaneswar/> accessed 30 November 2017

Bandopadhaya, R. (2009) "Hawker's Movement in Kolkata 1975–2007", *Economic and Political Weekly*, 44, 17, 116–19

BDA-BMC (2015) *Bhubaneswar Smart City Proposal*, Ministry of Housing and Urban Development, Government of Odisha, Bhubaneswar

BSCL (2015) *Smart City Plan*, Bhubaneswar Smart City Ltd <www.smartcitybhubaneswar.gov.in/smartcityFeatures> accessed 1 December 2016

BSCL (2017) *Towards a Smarter BBSR*, Bhubaneswar Smart City Ltd. < www.smartcityb hubaneswar.gov.in/project/33/0> accessed 24 December 2017

Centre for Civil Society (2015) *Assessment of State implementation of Business Reforms*, Centre for Civil Society <http://easeofdoingbusiness.org/state-rankings> accessed 5 October 2017

Chatterji, T. (2013) *Local Mediation of Global Forces in Transformation of the Urban Fringe*, Lambert Academic Publishing, Saarbrucken

Chatterji, T. (2017) "Modes of Governance and Local Economic Development: An Integrated Framework for Comparative Analysis of the Globalizing Cities of India", *Urban Affairs Review*, 53, 6, 955–89

Datta, A. (2015) "New Urban Utopias of Postcolonial India: 'Entrepreneurial Urbanization' in Dholera Smart City, Gujarat", *Dialogues in Human Geography*, 5, 1, 3–22

Government of India (2011) *Population Census 2001–11*, Registrar General and Census Commissioner, Government of India, New Delhi

Kalia, R. (1995) *Bhubaneswar: From a Temple Town to a Capital City*, SIU Press, Carbondale

Ministry of Finance (2017) *Economic Survey 2016–17*, Government of India, New Delhi

Ministry of Urban Development (2017) *Smart Cities Mission*, Government of India <http://smartcities.gov.in/> accessed 12 July 2017

MoSPI (2017) *Statistical Year Book of India*, Government of India, New Delhi

NASSCOM (2016) *Strategic Review 2015*, National Association of Software and Service Companies <www.nasscom.org/indian-itbpo-industry> accessed 4 September 2017

Praharaj, S., Han, J.H. and Hawken, S. (2018) "Towards the Right Model of Smart City Governance in India", *International Journal of Sustainable Development and Planning*, 13, 2 171–86

Roy, S. (2016) "The Smart City Paradigm in India: Issues and Challenges of Sustainability and Inclusiveness", *Urban Affairs Review*, 44, 5–6, 29–48

Smart Cities Council (2018) "Pundits Call for Strategic Approach to Make Bhubaneswar e-waste Free" <https://india.smartcitiescouncil.com/article/pundits-call-strategic-approach-make-bhubaneswar-e-waste-free> accessed 24 January 2018

The Telegraph (2017) "Smart Vendors on Janpath", Ananda Bazar Patrika Group, <https://www.telegraphindia.com/1170228/jsp/odisha/story_138057.jsp> accessed 27 February 2017

12

SAN FRANCISCO, UNITED STATES

A new model of sustainable industrial land use

Carl Grodach and Declan Martin

Introduction

This chapter examines the sustainable economic development implications of the City of San Francisco's recent industrial planning efforts. We evaluate the extent to which the City's innovations in zoning and related industrial support efforts are able to help balance the needs of different economies and workforces and the complex land use and economic development trade-offs involved in a high-cost, high-demand city. The now familiar story of San Francisco's gentrification and "post-industrial" restructuring has resulted in a city of high-value office, residential, and consumption spaces, often by transforming industrial lands considered obsolete and underdeveloped (Chapple et al. 2017; Solnit and Schwartzenberg 2000). The production of urban space for advanced services, high technology, and commercial creative industries has enhanced real estate values without broad-based benefit (Hartman and Carnochan 2002; Stehlin 2016). Indeed, it has come at the cost of uneven development, pricing out low- and middle-income households as well as businesses, reducing opportunities for career mobility and access to living wage jobs.

Despite San Francisco's industrial gentrification, some established manufacturing operations have remained by adapting to shifting economic conditions. At the same time, new forms of urban manufacture are on the rise. This includes the "maker movement," which blends traditional craft with high-technology processes in niche production (Hatch 2013; Wolf-Powers et al. 2017). Advocates proclaim that the resurgence of urban manufacturing has the potential to create relatively high-paying, career ladder jobs while offering opportunities for sustainable resource use and "green" industry development (Langdon and Lehrman 2012; SFMade et al. 2016). However, the high-cost real estate market combined with the loss of suitable production space significantly harms the potential for development of an interconnected urban manufacturing sector to deliver its economic development potential.

San Francisco is amongst the forerunners to face the sustainability challenge of reintegrating a production base into the urban core. This effort may tackle sustainable economic development and social justice issues tied to "post-industrial" urban development by pursuing economic development planning geared toward diversifying the economic base and creating living wage jobs rather than boosting land values and attracting high-wage industries. The City is experimenting with new zoning measures and incentive schemes to protect the urban industrial base and balance the needs of urban manufacturers with the demand for residential and office space by channelling private investment into the development of new multiuse industrial spaces. The aim of this chapter is to examine whether such regulation creates the conditions for more sustainable economic development and land use options or continues the trend toward corporate upscaling of the city's already scarce industrial lands.

The City: San Francisco

San Francisco is the thirteenth largest city in the US with a 2016 population of 870,877. It comprises the major centre of the San Francisco-Oakland-Hayward, CA Metropolitan Statistical Area (4.6 million population) along with San Jose and the Silicon Valley to the south and the historically industrial, working class, and gentrifying East Bay communities (US Census Bureau, n.d.) (Figure 12.1). The city contains a dense urban core and compact mixed-use neighborhoods. At 6,266 people per square mile, it is the second densest urban area in the US behind Los Angeles (Cox 2014). It is also the wealthiest urban area in the US, with a 2016 median household income just under US$97,000 (Kopf and Varathan 2017), and possesses one of the most highly educated populations in the country (*New York Times* 2012). The Bay Area's gross metropolitan economic output of $575 million puts it on par with Argentina and Sweden (Storper 2013), yet the city rivals some developing nations in terms of income inequality (Knight 2014). San Francisco has one of the highest concentrations of financial, business, and legal services in the country (Sassen 2001) and is a hub of high-tech start-ups and venture capital (Florida and King 2016). It is also home to over 600 manufacturers that employ a majority of people from low-income households (SFMade et al. 2016).

San Francisco is the only consolidated city-county in the State of California. The San Francisco government derives its jurisdiction from the city charter and is comprised of an executive and legislative branch, the latter responsible for passing laws and appointing the Planning Commission who advise on growth and development in San Francisco (City and County of San Francisco 2006). The Office of Community Investment and Infrastructure (formerly San Francisco Redevelopment Agency) is responsible for the formation of public-private partnerships that deliver some of the largest redevelopment projects in the city and the Office of Economic and Workforce Development (OEWD) concentrates on business attraction and retention and workforce programs.

FIGURE 12.1 San Francisco Bay Area

Source: Public domain image: https://en.wikipedia.org/wiki/San_Francisco_Bay_Area#/media/File: Bayarea_map.png

San Francisco's industrial strategy must be placed not only in this govern-ance context, but also in the context of its urban development history. Despite its politically progressive population, the city form has been strongly influ-enced by a pro-growth coalition (Hartman and Carnochan 2002). Since the 1950s, development interests have induced commercial office development in the urban core as a replacement for declining industrial activity. As the office market declined in the 1980s, the growth coalition turned toward remaking former industrial zones like South of Market Area (SoMA) into arts and tour-ism destinations through flagship cultural projects and entertainment complexes

(Grodach 2010). The gentrification, particularly of SoMA, but also other central districts continued through the "dot-com" boom of the 1990s. That decade saw a significant amount of technology start-ups, multimedia firms, and residential developers adapt the area's industrial buildings to suit their needs (Graham and Guy 2002). This issue endured following the 2001 market crash and became more pronounced in 2011 after the City passed the Central Market Payroll Tax Exclusion Ordinance in response to Twitter's threat to re-locate their San Francisco headquarters (Warburg 2014).[1] Over this period, San Francisco's industrial zoned land, classified as Production, Distribution, and Repair (PDR), steadily shrank to around 12 per cent of the city's usable land (San Francisco Planning Department et al. 2014, 32).

This creative destruction of the built environment has resulted in a mixed-use "post-industrial" cityscape built around new forms of production – office, design and research and development – alongside residential and consumption areas. While this approach has delivered a robust tax base and significant employment to the city, it has also created major issues with congestion, pollution, rising rents and property values, the displacement of existing businesses and residents, and the destruction of the city's older office and industrial spaces (Godfrey 1997). San Francisco's slow-growth, anti-development opposition certainly reshaped some plans for development, but its focus on quality of life issues hardly addressed the preservation of industrial lands and working-class jobs (McGovern 1998). There are now serious concerns that the supply of industrial land in San Francisco is not sufficient to meet future manufacturing needs, with recent estimates indicating a vacancy rate of 3 per cent and projecting a deficit of industrial land by 2040 (Chapple et al. 2017, 29). A recent revival of production activity in San Francisco has been met with optimism by the San Francisco Government and Planning Department. However, policymakers face the complex challenge of developing regulations that effectively protect PDR zones and channel private investment into San Francisco's industrial facilities if an emerging urban manufacturing sector is to deliver on its promise of accessible middle-wage employment and resilient local economies.

Sustainable economic development is distinguished from conventional approaches in that it considers economic, environmental, and equity impacts together rather than prioritizing economic growth (Fitzgerald and Green Leigh 2002; Grodach 2011). In this regard, San Francisco – like many cities – now faces the sustainability challenge of addressing its "post-industrial" urban development program, focused predominately on enhancing land values at the expense of equitable development. Policy geared toward remaking industrial lands for technology, business services, and tourism has facilitated the rapid upscaling of San Francisco's older warehouse and industrial buildings and resulted in a loss of living-wage jobs and affordable housing. Urban manufacturers and other small-to-medium enterprises (SMEs), an important source of such employment, have been subjected to displacement pressures as a result of the shrinking stock of central industrial real estate and competition from other sectors able to pay higher rents. In 2014, industrial rents in San Francisco proper averaged $26.26 per square foot, significantly higher than the regional average, and industrial rents in SoMA averaged $41.53 per square foot (Chapple et al. 2017, 33).

San Francisco's prohibitive real estate market has posed significant obstacles for manufacturing in the region. This is especially problematic because nationwide manufacturing has experienced relatively steady growth since the Global Financial Crisis and, in some places, outperformed the national average in employment gains (Christopherson 2011). These jobs provide relatively high wages, employee health care benefits, and upskill opportunities for workers with lower educational attainment (Langdon and Lehrman 2012). Despite the Bay Area's high-cost real estate market, manufacturing accounted for approximately 11,000 jobs in 2015 and created approximately $614 million in direct sales for the regional economy (SFMade et al. 2016, 14). This has elicited support for re-industrialization strategies, particularly in regions such as San Francisco with a comparative advantage in SME-driven advanced manufacturing and the production of consumer goods due to the region's history of computer and electronics research, design, and manufacturing along with its established entrepreneurial system (SFMade et al. 2016). In fact, manufacturing firms in the city employ on average 15 people (SFMade et al. 2016, 14) and operate as part of a vertically disintegrated chain of production that taps into the city's urban infrastructure, skilled labor, and business networks.

The San Francisco Government is conscious of the city's strong urban manufacturing ecosystem, but faces complex challenges in preserving it. The first challenge is maintaining the city's scarce stock of industrial land in strategic locations that are proximate to key infrastructure and labor supply. These industrial sites face considerable pressure for rezoning and upscaling to mixed-use office and residential particularly from the burgeoning technology sector. This is further compounded by a deficit in affordable central city housing (San Francisco Planning Department 2016). Rezoning of San Francisco's few remaining industrial strongholds would likely have a permanent and irreversible effect on the ability of urban manufacturers to operate in the city, precipitating an unsustainable increase in rents from competing uses and forcing most manufacturing SMEs to emigrate or cease operations.

The second challenge is incentivizing the construction of new industrial space and maintaining and upgrading existing space. While demand for industrial space in San Francisco is considerable (CBRE 2017), the capital outlays required to construct new space are highly disproportionate to the rents that can be achieved from other uses. For instance, the average asking price for office rent was roughly 2.5 times higher per square foot annually than industrial rents in 2014.[2] Furthermore, San Francisco's industrial buildings are typically between 30 and 100 years old, with many requiring major capital outlays for upgrades to make them functional production spaces (San Francisco Planning Department et al. 2014, 34). A recent study of San Francisco's food and beverage manufacturers estimated a capital outlay of $250,000 to ensure adequate facilities and regulatory compliance of a 5,000 square foot production and distribution facility before operations could begin (San Francisco Planning Department et al. 2014, 33). The cost of redevelopment is compounded by the city's large, single-occupant industrial building stock. These buildings are usually several times the requisite size for San Francisco's SME manufacturers, generating additional costs and regulatory complexity to create sub-divided shared facilities (SFMade et al. 2016, 24). Moreover, the capital outlays and complex regulatory environment for industrial activity creates substantial risk for developers and investors relative to more passive uses of industrial buildings such as storage (Urban Manufacturing Alliance 2014).

Taken together, the lower risk attached to passive uses and the greater financial return on office and residential development creates a strong disincentive to pre-serve, upgrade, and produce new industrial facilities. Given the strong competition for central, strategically located land in the urban core, policymakers are faced with complex trade-offs to balance different land uses and adequately incentivize the provision of modern industrial space that caters to the needs of contemporary urban manufacturing.

The planning innovation

San Francisco's planning innovation revolves around 1) the revision and enforce-ment of its PDR zoning code to protect urban industrial lands and limit competing uses, 2) zoning variances and incentives aimed at leveraging real estate demand to increase the supply of industrial space, and 3) an industrial rebranding campaign led by non-profit industrial advocacy group, SFMade. These adaptations were solidi-fied and promoted in former Mayor Ed Lee's five-point Plan for Manufacturing, which aimed to preserve existing industrial space, incentivize the development of new PDR space on private land, and build and upgrade space on public lands (San Francisco Office of Economic and Workforce Development n.d.).

The innovation process

In 2002, the City imposed temporary controls in response to land use con-flicts between residential and non-residential uses, precipitating a prolonged period of public consultation to develop a comprehensive plan for the Eastern Neighborhoods including East SoMA, the Mission, Showplace Square, Potrero Hill, Dogpatch, and Central Waterfront areas (San Francisco Planning Department 2007, S1) (Figure 12.2). The plan's main aim was to identify appropriate locations for housing in the City's industrial zones and to address problems around affordable housing. However, a vocal industrial constituency steered the debate toward retention of an adequate supply of industrial land (Mission Coalition for Economic Justice and Jobs 2003; San Francisco Planning Department 2007). The final plan attempted to strike a balance by allowing the transition of some industrially zoned land toward residential and mixed use, but tightened controls prohibiting non-PDR uses in the more traditional industrial zones. The overall impact was a reduction in the amount of industrial land in the Eastern Neighborhoods with more significant protections for the remaining

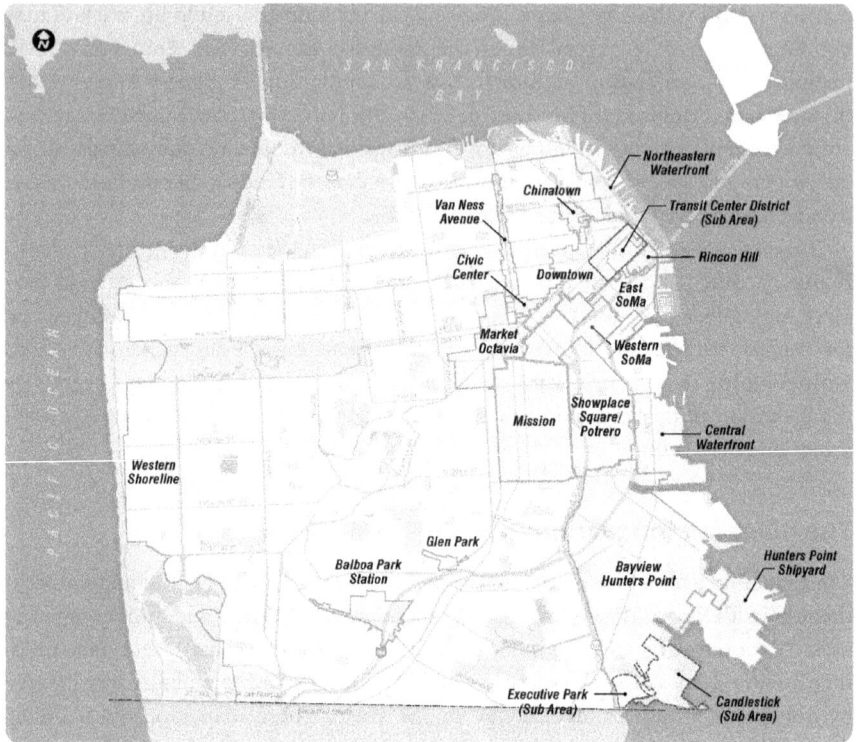

FIGURE 12.2 Eastern Neighborhoods Plan areas

Source: San Francisco Planning Department (n.d.)

PDR zones to mitigate real estate speculation and ad hoc development of non-conforming uses in industrial areas (Economic and Planning Systems 2005; San Francisco Planning Department 2007). In 2008, the City adopted the Eastern Neighborhoods Plan that established PDR zones, preserving approximately 7 per cent of San Francisco's total usable land for PDR businesses (San Francisco Planning Department et al. 2014, 31-32).

The adoption of these ordinances was the result of a protracted debate in the aftermath of the dot-com boom around gentrification and land use in San Francisco's last remaining industrial concentration in the southeastern portion of the city. The rapid expansion of "post-industrial" development throughout the 1990s and early 2000s significantly altered the city's land use system, precipitating industrial displacement. Advocacy groups including the Mission Coalition for Economic Justice and Jobs and the South of Market Community Action Network along with residents, workers, and business owners in San Francisco's industrial pockets engaged with City planners in negotiating a solution. Drawing on the decline in PDR jobs throughout the latter twentieth century and entrenched assumptions about the incompatibility of industry within the central city, opponents of PDR districts argued for the use of industrial land to meet the city's housing and office needs (Wertheim 2015). In response, manufacturing advocates built a case around the importance of industrial land in providing for physical infrastructure industries (e.g. construction, warehousing, storage, repair, and manufacturing) that are unable to compete for space in an unregulated market (San Francisco Planning Department 2002; San Francisco Office of Economic and Workforce Development 2014, 10). They argued that these industries provide essential services that are not amenable to outsourcing and support the growth of knowledge and tourism industries, while generating strong economic multipliers and substantial employment for workers without high educational attainment (San Francisco Office of Economic and Workforce Development 2014, 10; Economic and Planning Systems 2005).

However, the protections afforded to manufacturing and its subsequent resurgence in San Francisco may have been more the product of serendipity than astute planning. According to Steve Wertheim (2015), a project manager involved with the Eastern Neighborhoods Plan, "at the time [the Planning Department] created the PDR Districts, [they] weren't doing so to protect [the City's] manufacturing sector, which seemed beyond saving." Nevertheless, the creation of these zones created affordable land in strategic urban locations sheltered from incompatible and higher value uses, which provided space for a range of emerging SME manufacturers. These manufacturers differed from physical infrastructure industries that the PDR districts were crafted for: they utilized advanced technologies and flexible production methods to create market-oriented products including medical devices, apparel, and food and beverages, rather than services such as storage, construction, and repair (SF Made et al. 2016).

Advanced and consumer-product manufacturers benefitted from the geographical concentration of consumers, competitors, and suppliers to facilitate rapid just-in-time turnarounds, as well as proximity to freight services, labour, and the San Francisco "urban brand" (SF Made et al. 2016; San Francisco Planning Department et al. 2014; Maskell 2001). The emergence of such manufacturers coincided with the increasing momentum of the Maker Movement in San Francisco as well as national policy efforts aimed at stimulating advanced manufacturing for employment and economic growth opportunities (Friedman and Byron 2012; Hatch 2013).

In San Francisco, this was carried forward by manufacturing intermediary SFMade. The non-profit emerged as a key intermediary in representing the needs and collective role of local manufacturers in 2010. Initiated by a local manufacturer of handbags to promote and develop small manufacturing in the city, SFMade is now funded in part by and works in partnership with the City. SFMade has been instrumental in rebranding local manufacturing as an innovative and entrepreneurial growth sector, emphasizing its interdependence with San Francisco's technology and arts ecosystems and its capacity to develop new products, services, and business models (SFMade et al. 2016, 19–23; San Francisco Planning Department et al. 2014, 12). In this way, it aims to rethink the entrenched "post-industrial" policy narrative, presenting an alternative to tech and real estate driven development by fulfilling workforce development, real estate, and advocacy functions to foster an expanded manufacturing base.

In 2014, SFMade collaborated in a process of legislative reform with the Mayor's Office, Supervisor's Office, and Planning Department to incentivize the preservation, upgrading and construction of space for manufacturers (San Francisco Planning Department 2014, 3). Despite the protections afforded by the PDR districts in the Eastern Neighborhoods Plan, manufacturers still faced issues with an aging industrial building stock comprised of predominately large single-occupancy facilities. As a result, planners attempted to leverage market demand for higher paying office uses to create new PDR space primarily through so-called hybrid development projects. To do so, the City amended PDR zoning in 2014 to permit new office construction larger than 20,000 square feet on vacant or near-vacant land around the Eastern Neighborhoods, provided at least 33 per cent of the gross square footage is maintained for PDR use (San Francisco Planning Department, 2014, 19–20). The Planning Code Section 219.1 proactively seeks to incentivize the construction of new PDR facilities by requiring higher paying office uses to subsidize development.

One Hundred Hooper Street is the first project proposed under the new Code. The four-story mixed-use project in Potrero Hill/Dogpatch will include approximately 86,000 square feet of PDR space and 284,000 square feet of office space across three buildings on a former self-storage site adjacent to California College of the Arts (San Francisco Planning Department 2015). Located in an already strong design cluster around the college, the project may promote synergies around design-oriented manufacturing. Further, the developer is working in partnership with the College and PlaceMade, an affiliate of SFMade specializing in non-profit industrial property development. PlaceMade, which will handle leasing of the "PDR workshop" building, is a critical intermediary between prospective industrial tenants, the developer, and city planners that helps to ensure the needs of manufacturers are met. The project is expected to generate up to 450 manufacturing jobs (San Francisco Planning Department 2015).

Following Hooper Street, proposals have included new build industrial buildings and mixed-use PDR and office space in the historic Armory in the Mission District and mixed PDR, arts, and maker spaces in Hunter's Point Shipyard and Forest City's Pier 70 in an attempt to speed up approval processes (McKinnon 2016). Additionally, in Central SoMA, a draft plan is currently under public review that recommends mechanisms to provide for new PDR space in arts and light industrial zones undergoing rezoning to mixed-use office, involving provisions for on- and off-site replacement, the potential for an in-lieu fee to the City for new PDR construction and preservation, and transferable development rights (TDRs) on important industrial sites for use on another development (San Francisco Planning Department 2016, 109, 145). Alongside these zoning innovations, the City continues to provide conventional site selection, business development, and zoning and permitting assistance for PDR space.

Trade-offs

Due to the relatively recent implementation of this legislation, the implications and trade-offs can only be surmised. The cross-subsidy, replacement, and TDR mechanisms provide an innovative market-based solution to the loss of and under-investment in urban industrial space. However, they are also a response to the on-going pressures from real estate development interests, which have long helped to shape the city's pattern of urban development including the loss of industrial lands. Additionally, without stringent on-going monitoring of mixed-use office and industrial projects, this legislation could further expose the city's remaining industrial pockets to irreversible commercial upscaling to higher paying uses. This would not only create additional harm to the development of an urban manufacturing base, but also reduce the ability to deliver affordable housing in an area in need.

While San Francisco's hybrid development model and PDR replacement provisions represent an innovative means of channeling private investment into industrial development, it remains to be seen whether such investment can be shaped in a way that prioritizes the needs of the city's urban manufacturing and employment base. The first consideration is whether developers will take advantage of the broad spectrum of PDR businesses and select tenants on the basis of their compatibility with the non-PDR dimensions of the development. For instance, the Central SoMA Strategy encourages the use of makerspaces as a ground floor activator combined with retail and organized cultural events to engage pedestrians and maintain SoMA as a "place of production" (San Francisco Planning Department 2016, 155). Although makerspaces provide affordable and shared access to equipment and technical knowledge, they do not address the real estate needs of operational PDR businesses, which are more likely to contribute to manufacturing supply chains and employment (Wolf-Powers et al. 2017). PDR businesses that have special space needs around drainage, ventilation, and loading access require more significant capital outlays, making them less appealing to developers of hybrid precincts. Food and beverage production, for example, requires the installation of oven hoods, grease traps, additional drainage and loading docks to comply with planning regulations and the public health code, which can cost approximately $50 per square foot (San Francisco Planning Department et al. 2014, 33). This cost presents a disincentive for developers to incorporate such space, despite its sizeable and growing share of living-wage production jobs and strong space demand (San Francisco Planning Department et al. 2014). Without adequate regulation to correct incentives, the capacity of hybrid development projects to provide PDR space to strengthen urban manufacturing activity and employment may be jeopardized.

A second consideration pertains to the proposed provisions that allow the off-site replacement of PDR land in arts and light industrial districts that have been rezoned for mixed-use office space. While this provision offers flexibility in meeting requirements for PDR preservation and construction and the potential to produce more space for PDR jobs, some sites may have locational attributes that cannot be easily replicated (Lester et al. 2013). Sites that facilitate inter-relationships with local markets and retailers, access to key transportation infrastructure, and the development of a distinct local production culture and brand are typically path dependent and cannot simply be uprooted to another location (Gibson 2015).[3] Furthermore, locational requirements differ significantly between PDR businesses, ranging from publishing and interior design to construction and auto repair. For example, craft and cultural product manufacturers generally benefit from vibrant areas that enable cross-fertilization between sectors and possess a strong image or historic legacy that can be used in marketing and design (Fox Miller 2017; Rantisi and Leslie 2010; Scott 1996). As such, these organizations predominately locate in the central city light industrial areas of SoMA and Showplace Square and are at risk of displacement if PDR land conversions in these areas are not replaced with similar, centrally located light industrial space with a complementary mix of uses and infrastructure (Economic and Planning Systems 2005, 42). A finely tuned approach is required that replaces converted industrial land with new PDR space in strategic locations for organizations that strengthen the city's production system and provide employment (Howland 2010). This is a particular challenge in a high-rent city facing pressures on affordable housing as well as production space.

Finally, the success of hybrid office-industrial precincts and PDR replacement provisions is highly contingent on tight controls and monitoring of private development. Mixed-use PDR developments are reviewed by the Planning Commission in an attempt to "maximize the potential for . . . project[s] to produce new PDR space that is viable and affordable" and includes provisions for continued reporting and monitoring to ensure the prioritization of PDR uses on project sites (San Francisco Planning Department 2015, 6). However, the City has demonstrated leniency toward ad hoc office conversions in the past (Bronstein 2016; Dineen 2016). Moreover, PDR zoning is not without loopholes. Notably, historic landmark buildings are exempt from PDR restrictions, enabling development of office space in industrial areas (Bronstein 2014). Without adequate supervision and enforcement to ensure the ongoing compatibility of PDR and non-PDR uses, the City's planning innovations are at risk of opening the city's scarce supply of industrial land to upscale development. And, the reality is that such projects will only move forward if developers can achieve their return on investment. In short, while the move toward industrial mixed-use development is positive, this market-led approach will face challenges around equitable development and the wider issue of delivering both industrial space and affordable housing in a high-rent property market.

Conclusion

San Francisco's strategy for the preservation, maintenance, and construction of industrial space has emerged as a geographically and historically contingent process. The rise of an advanced service and tech economy precipitated the construction of offices, upscale apartments, and retail and entertainment precincts, marginalizing established industrial and working-class uses in the city's Eastern Neighborhoods. The pace and magnitude of this "post-industrial" transition rapidly set the city on a path of service-dependent development that was only seriously reflected upon in the aftermath of the dot-com crash. What initially began as a means of providing space for San Francisco's physical infrastructure industries gradually evolved into a comprehensive urban manufacturing land use strategy.

The design and implementation of this policy strategy is a continuing process, spanning over decades and involving a number of propitious place-specific circumstances. San Francisco forms a vital node in the regional Bay Area economy, characterized by an embedded entrepreneurial culture sustained through venture capital and incubation infrastructure, design and engineering expertise, and a strong artist and maker community. These factors have contributed to a resurgence of manufacturing, driving a corresponding increase in demand for industrial land. In conjunction, the emergence of key intermediaries, particularly SFMade, but also the community groups that fought the loss of industrial land through the dot-com gentrification period, have been crucial in developing the new thinking around San Francisco manufacturing. Against the entrenched narrative of post-industrialism, SFMade in particular has rebranded manufacturing to policymakers as crucial to local identity and as a viable source of living-wage employment and sustainable economic development. Their advocacy work, along with that of Eastern Neighborhood community groups, encouraged the City to preserve industrial land in a high-cost, high-demand property market.

In an environment of intensified and instantaneous connectivity where policy rapidly traverses different geographic sites, it is important to acknowledge that San Francisco's industrial land use strategy emerged in part as a result of these local processes and actors. Without the City's history of community advocacy, informed planners, and SFMade, the industrial lands strategy likely would not have occurred. Still, while the process cannot be seamlessly emulated across cities, its basic components are certainly applicable in other places with reemerging manufacturing economies and a scarce supply of industrial land. Public entities in other gentrified post-industrial cities including the Boston Planning and Development Agency (2017), New York City Council (2014), and City of Portland (2015), have looked to San Francisco as a model, exploring and implementing similar zoning strategies. Nonetheless, San Francisco and cities like it will continue to face the challenge of balancing the needs of production activity with competing uses and other important agendas, particularly affordable housing, as they wrestle with zoning tools geared toward market demand.

Notes

1 The ordinance waived six years of payroll taxes for organizations with a payroll exceeding $250,000 provided they set up in the Central Market Street and Tenderloin area, establishing a technology hub in an area that has struggled with under-investment, physical blight, and poverty for several decades (Stehlin 2016).

2 Due to the relatively limited data on industrial rents in San Francisco proper, a rough estimate was produced to illustrate the magnitude of the rent gap between industrial and office uses, based on the average industrial rent of $26.26 for 2014 presented by SFMade (SFMade et al. 2016, 23) and the year-end average office rent for 2014 presented by Colliers International (Colliers International 2014, 5).

3 Locational attributes may not be pertinent considerations for the Central SoMA Plan proposal, which requires the replacement of PDR land at another location within SoMA (San Francisco Planning Department 2016, 109). However, the proposition of an in-lieu fee to the City for PDR maintenance and construction will require strategic deliberations when deciding on where to reinvest.

References

Boston Planning and Development Agency (2017) "Raymond Flynn Marine Park master plan update" (www.bostonplans.org/planning/planning-initiatives/raymond-l-flynn-marine-park-master-plan-update) accessed 27 November 2017

Bronstein, Z. (2014) "Why SF City Planning can't protect local industry from office encroachment: An alarming case study", *48 Hills*, 29 May (https://48hills.org/2014/05/sf-city-planning-cant-protect-local-industry-office-encroachment-alarming-case-study/) accessed 27 April 2018

Bronstein, Z. (2016) "Illegal office conversion: A case study", *48 Hills*, 17 June (https://48hills.org/2016/06/illegal-office-conversion-case-study/) accessed 1 March 2018

CBRE (2017) "San Francisco Bay Area industrial market view snapshot Q1 2017" (https://www.cbre.com/research-and-reports/San-Francisco-Bay-Area-Industrial-MarketView-Snapshot-Q1-2017) accessed 1 March 2018

Chapple, K., Ritter, S., Ross, A., Mattuizzi, E., Lapeyrolerie, E., and St-Louis, E. (2017) "Industrial land and jobs study for the San Francisco bay area: Final report", (CA17-2792) University of California, Berkeley (www.planningfor.jobs/sites/default/files/memo_1_3.15.pdf) accessed 1 March 2018

Christopherson, S. (2011) "Riding the small wave in manufacturing to more good jobs and a more diverse economy", Big Ideas for Job Creation in a Jobless Recovery, The Institute for Labor and Employment, UC Berkeley (www.bigideasforjobs.org/wp-content/uploads/2011/09/Christopherson-Full-Report-PDF1.pdf) accessed 1 March 2018

City and County of San Francisco (2006) "City and County of San Francisco municipal code: 1996 charter" (https://archive.org/stream/gov.ca.sf.charter/ca_sf_charter - page/n0/mode/2up) accessed 1 March 2018

City of Portland (2015) "Central City 2035 Southeast Quadrant Plan" (https://www.portlandoregon.gov/bps/68508) accessed 1 March 2018

Colliers International (2014) "2014 year-end and forecast report: San Francisco" (www.colliers.com/-/media/files/united states/markets/san francisco/marketreports/2014yearend_sanfranciso.pdf?la=en-US) accessed 1 March 2018

Cox, W. (2014) "America's densest cities", *Huffington Post*, September 26 (https://www.huffingtonpost.com/wendell-cox/americas-densest-cities_b_5888424.html) accessed 1 March 2018

Dineen, J. K. (2016) "Offices intruding on SF space zoned for industrial use", *San Francisco Chronicle*, 14 March (www.sfchronicle.com/bayarea/article/Offices-intruding-on-SF-space-zoned-for-6889809.php) accessed 1 March 2018

Economic and Planning Systems (2005) "Supply/demand study for production, distribution, and repair (PDR) in San Francisco's Eastern Neighborhoods", (http://sf-planning.org/sites/default/files/FileCenter/Documents/1929-14158FinRpt1.pdf) accessed 1 March 2018

Fitzgerald, J. and Green Leigh, N. (2002) *Economic Revitalization: Cases and strategies for city and suburb*. Sage, Thousand Oaks

Florida, R. and King, K. (2016) *Rise of the Global Startup: The geography of venture capital investment in cities and metros across the globe*. Martin Prosperity Institute, Toronto

Fox Miller, C. (2017) "The contemporary geographies of craft-based manufacturing", *Geography Compass*, 11, 4, 1–13

Friedman, A. and Byron, J. (2012) "High-tech, high-touch, and manufacturing's triple bottom line", *Innovations*, 7, 3, 83–95

Gibson, C. (2015) "Material inheritances: How place, materiality, and labor process underpin the path-dependent evolution of contemporary craft production", *Economic Geography*, 92, 1, 61–86

Godfrey, B. J. (1997) "Urban development and redevelopment in San Francisco", *Geographical Review*, 87, 3, 309–333

Graham, S. and Guy, S. (2002) "Digital space meets urban place: Sociotechnologies of urban restructuring in downtown San Francisco", *City*, 6, 3, 369–382

Grodach, C. (2010) "Beyond Bilbao: Rethinking flagship cultural development and planning in three California cities", *Journal of Planning Education and Research*, 29, 3, 353–366

Grodach, C. (2011) "Barriers to sustainable economic development: The Dallas-Fort-Worth experience", *Cities*, 28, 4, 300–309

Hartman, C. and Carnochan, S. (2002) *City for Sale: The transformation of San Francisco*. University of California Press, Berkeley and Los Angeles

Hatch, M. (2013) *The Maker Movement Manifesto: Rules for innovation in the new world of crafters, hackers, and tinkerers*. McGraw-Hill Education, New York

Howland, M. (2010) "Planning for industry in a post-industrial world", *Journal of the American Planning Association*, 77, 1, 39–53

Knight, H. (2014) "Income inequality on par with developing nations", *SF Gate*, 25 June (https://www.sfgate.com/bayarea/article/Income-inequality-on-par-with-developing-nations-5486434.php) accessed 1 March 2018

Kopf, D. and Varathan, P. (2017) "Boomtown by the bay. It's official: San Francisco is the richest urban area in America", *Quartz*, September 14 (https://qz.com/1077050/san-francisco-is-americas-richest-major-metropolitan-area-rising-above-washington-dc-in-the-latest-rankings/) accessed 1 March 2018

Langdon, D. and Lehrman, R. (2012) "The benefits of manufacturing jobs" (http://www.esa.doc.gov/sites/default/files/1thebenefitsofmanufacturingjobsfinal5912.pdf) accessed 1 March 2018

Lester, T. W., Kaza, N., and Kirk, S. (2013) "Making room for manufacturing: Understanding industrial land conversion in cities", *Journal of the American Planning Association*, 79, 4, 295–313

Maskell, P. (2001) "Towards of knowledge-based theory of the geographical cluster", *Industrial & Corporate Change*, 10, 4, 921–943

McGovern, S. (1998) *The Politics of Downtown Development: Dynamic political cultures in San Francisco and Washington D.C.* The University Press of Kentucky, Lexington

McKinnon, S. (2016) "PDR 101 (Production, Distribution, Repair)", *Arts for a Better Bay Area*, 27 January (https://www.betterbayarea.org/pdr_101) accessed 1 March 2018

Mission Coalition for Economic Justice and Jobs (2003) "An alternative future for the north east mission industrial zone" (https://www.yumpu.com/en/document/view/18971807/an-alternative-future-for-the-north-east-mission-industrial-zone) accessed 1 March 2018

New York City Council 2014 "Engines of opportunity: Reinvigorating New York City's manufacturing zones for the 21st century" (http://167.153.240.175/downloads/pdf/NYEO.pdf) accessed 1 March 2018

New York Times (2012) "Cities with the most college-educated residents", *New York Times* (www.nytimes.com/interactive/2012/05/31/us/education-in-metro-areas.html) accessed 1 March 2018

Rantisi, N. and Leslie, D. (2010) "Materiality and creative production: The case of the Mile End neighborhood in Montréal", *Environment and Planning A*, 42, 12, 2824–2841

San Francisco Office of Economic and Workforce Development (2014) "San Francisco economic strategy: 2014 update" (http://oewd.org/sites/default/files/FileCenter/Documents/613-FINAL_Economic Strategy 2014 Update 3.13.15.pdf) accessed 1 March 2018

San Francisco Office of Economic and Workforce Development (n.d.) "Mayor Lees' 5-Point Plan for PDR" (http://oewd.org/Industrial) accessed 1 March 2018

San Francisco Planning Department (2002) "Industrial land in San Francisco: Understanding production, distribution and repair" (http://sf-planning.org/sites/default/files/FileCenter/Documents/4893-CW_DPR_chapter5_2.pdf.) accessed 1 March 2018

San Francisco Planning Department (2007) "Eastern Neighbourhoods rezoning and area plans: Draft environmental impact report" (http://sf-planning.org/sites/default/files/FileCenter/Documents/3965-EN_DEIR_Part-1_Intro-Sum.pdf.) accessed 1 March 2018

San Francisco Planning Department (2014) "Planning code text change: PDR facilitation" (http://commissions.sfplanning.org/cpcpackets/2013.1896T.pdf) accessed 1 March 2018

San Francisco Planning Department (2015) "Conditional use authorization, planned unit development & office allocation" (http://commissions.sfplanning.org/cpcpackets/2012.0203BC.pdf) accessed 1 March 2018

San Francisco Planning Department (2016) "Central SOMA plan and implementation strategy" (http://default.sfplanning.org/Citywide/Central_Corridor/Central_SoMa_Plan_Part01-Central_SoMa_Plan_FINAL.pdf) accessed 1 March 2018

San Francisco Planning Department (n.d.) "San Francisco general plan" (http://generalplan.sfplanning.org/index.htm) accessed 1 March 2018

San Francisco Planning Department, San Francisco Office of Economic and Workforce Development and SPUR (2014) "Makers and movers economic cluster strategy" (https://www.spur.org/sites/default/files/wysiwyg/u76/SF_Makers_Movers_Economic_Cluster_Strategy_Nov_2014_low_res.pdf) accessed 1 March 2018

Sassen, S. (2001) *The Global City: New York, London, Tokyo*. 2nd Ed. Princeton University Press, New Jersey

Scott, A. J. (1996) "The craft, fashion, and cultural-products industries of Los Angeles: Competitive dynamics and policy dilemmas in a multisectoral image-producing complex", *Annals of the Association of American Geographers*, 86, 2, 306–323

SFMade, Mayor's Office of Civic Innovation, and San Francisco Office of Economic and Workforce Development (2016) "Make to manufacture: Advanced manufacturing playbook" (www.sfmade.org/blog/make-to-manufacture-advanced-manufacturing-playbook/) accessed 1 March 2018

Solnit, R. and Schwartzenberg, S. (2000) *Hollow City: Gentrification and the eviction of urban culture*. Verso, London

Stehlin, J. (2016) "The post-industrial 'shop floor': emerging forms of gentrification in San Francisco's innovation economy", *Antipode*, 48, 2, 474–493

Storper, M. (2013) *Keys to the City: How economics, institutions, social interaction, and politics shape development*. Princeton University Press, New Jersey

Urban Manufacturing Alliance (2014) "Non-profit real estate development toolkit: Stable, affordable space for manufacturing" (http://urbanmfg.org/uma-content/uploads/2013/03/NonProfitRealEstateDevelopmentToolkitFINAL.pdf) accessed 1 March 2018

US Census Bureau (n.d.) "Community facts: San Francisco County, California" (https://factfinder.census.gov/faces/nav/jsf/pages/community_facts.xhtml) accessed 1 March 2018

Warburg, J. (2014) "Forecasting San Francisco's economic fortunes", SPUR, 27 February (www.spur.org/news/2014-02-27/forecasting-san-francisco-s-economic-fortunes) accessed 1 March 2018

Wertheim, S. (2015) "The history of zoning control in San Francisco", Urban Manufacturing Alliance, 21 July (www.urbanmfg.org/the-history-of-zoning-control-in-san-francisco) accessed 16 November 2016

Wolf-Powers, L., Doussard, M., Schrock, G., Heying, C., Eisenburger, M., and Marotta, S. (2017) "The maker movement and urban economic development", *Journal of the American Planning Association*, 83, 4, 365–276

13

ESCH-SUR-ALZETTE, LUXEMBOURG

The "Science City" in Belval – planning a large-scale urban development project in a small country

Tom Becker, Markus Hesse and Annick Leick

Introduction

The reconversion of large, centrally located brownfield sites in proximity to a growing city constitutes not only a great tool for the physical regeneration of a city but also a unique chance to support social and economic change towards more sustainable societies (cf. Jones and Evans 2013). Large-scale urban reconversion schemes such as the Belval project with the Cité des Sciences (Science City) quarter in Esch-sur-Alzette are by no means a new post-industrial phenomenon. Over the past decades, hundreds of projects all over the world have been designed and implemented with diverging results (Karadimitriou et al. 2013; Ponzini 2011; Taşan-Kok 2009; Gold and Gold 2008; Swyngedouw et al. 2002; Carrière and Demazière 2002; Moulaert et al. 2001). What makes the Luxembourg case specific, however, is the trans-scalar importance attributed to the project, the inventive implementation of the concept of sustainability and the innovative governance approach adopted for managing it.

Covering a total 120 hectares, the Belval reconversion programme is not only a flagship project for Esch-sur-Alzette, the Red Rock region or the Grand Duchy but also beyond. Given its extent and its innovative approach to sustainable growth, Belval constitutes a European model project (European Commission 2013). As other contributions in this book show, the urban context is generally too complex for the individual dimensions of sustainability to be delivered equally. Evaluations carried out on behalf of the development agencies involved indicate that, in the case of Belval, and more specifically the Cité des Sciences, economic innovation and environmental sustainability came at the expense of social cohesion (Belval 2011; Le Fonds Belval 2011a, 2011b). Governance is crucial in this respect. Whereas the selected governance approaches were certainly pioneering for Luxembourg at the time and may have contributed to the efficiency and effectiveness of the project execution, they were far from being truly integrated. From a sustainable governance perspective (Grin et al. 2010; Meadowcroft et al. 2005), the planning and implementation of Belval has led – and still leads – to the exclusion of important stakeholder groups. This, in turn, has substantial repercussions on the quality and acceptance of the project.

Our analysis builds on the one hand on insights from a PhD thesis on discourses, practices and social worlds in decision and planning processes in Luxembourg, including inter alia the Belval case (Leick 2016). The empirical basis for the latter comprises 20 in-depth interviews of experts of one to three hours each. Furthermore, a large number of documents, such as official reports of the development agencies or the ministries as well as documents of the national parliament, for instance the transcriptions of parliamentary debates, were analysed (Leick 2016). On the other hand, this chapter relies on long-standing experiences and debates from the 'Belval Observatory', a platform for exchange and knowledge gathering for a wide range of stakeholders active in research and practice on Belval between 2010 and 2015. The background information on Esch-sur-Alzette as well as on Belval and the Cité des Sciences stems from a desk research.

The city: Esch-sur-Alzette

Over the past two decades, the small city of Esch-sur-Alzette, situated on the edge of the Franco-Luxembourgian border in the so-called Red Rock region, has undergone a series of major transformations because of rapid global industrial restructuring. Once an industrial stronghold and home to the country's prime steel production sites, the city is now considered a notable symbol for socio-economic progress, not least because of the large-scale urban regeneration project of Belval. As a result of national efforts of strengthening the knowledge economy, the city has emerged as a knowledge city hosting the Cité des Sciences, a major research hub composed of the Grand Duchy's only university as well as several applied research institutions and incubators.

Esch-sur-Alzette is the second largest city in Luxembourg with approximately 35,000 inhabitants. Apart from a period of moderate decline between the 1970s and 1990s caused by the global economic downturn, the city has experienced a sustained population growth ever since the late nineteenth century, thus following similar trends in other small industrial cities like Differdange and Dudelange. As shown in Table 13.1, the population rose by 43.1 per cent between 1991 and 2017. Esch-sur-Alzette is the most densely populated city in Luxembourg at 2,395.5 people per square kilometre, with a land area of 14.35 square kilometres (Statec 2018). This high population density is primarily due to the city's compact historical inner-city neighbourhoods.

TABLE 13.1 Key characteristics of the four largest cities in Luxembourg

City	Number of inhabitants 1991	Number of inhabitants 2017	Variation 1991–2017 in %	Share of foreigners in % (2017)	Population density (inhabitants per km2) (2017)
City of Luxembourg	75,833	114,303	50.7	70.7	2,221.2
Esch-sur-Alzette	**24,018**	**34,378**	**43.1**	**57.2**	**2,395.7**
Differdange	16,096	25,402	57.8	55.9	1,145.3
Dudelange	14,254	20,480	43.7	40.9	957.9
Grand Duchy of Luxembourg	**384,634**	**590,667**	**53.6**	**47.6**	**228.4**

Source: Statec (2018)

The profound economic change, consisting of the diversification of the country's industry as well as the development of the tertiary sector, above all the financial centre, together with the resulting high growth rates that the Grand Duchy has experienced since the 1970s (Thewes 2017) have led to important socio-spatial reconfigurations of both the city and its population. Traditionally, its labour force was composed of Luxembourgers and high levels of migrants from Italy and Portugal who worked in low-skilled blue-collar jobs and lived in the city. Following the post-1970s' and post-2000s' state-led economic restructuring and diversification endeavours, several trends can be observed: first, the share of low-skilled cross-border commuters and migrant workers and artisans from Southern and Eastern Europe has increased. Second, the better-qualified Luxembourgish labour force tends to work in the service sector and commutes daily to and from Luxembourg City, situated 25 km north of Esch-sur-Alzette. Third, unemployment rates among the migrant work force are substantially higher than among Luxembourgish employees (ProSud 2017). Fourth, due to the establishment of the research hub in Belval in 2015, the share of well-educated and highly specialised researchers moving to Esch-sur-Alzette has risen considerably (Statec 2017; CEFIS 2013).

By implementing the Cité des Sciences/Belval, Esch-sur-Alzette ceased to be a self-contained and monocentric city surrounded by large industrial production sites. The decommissioning and subsequent reconversion of the former steel production site of Belval allowed for the development of a second city centre. In 2002, a master plan was conceived, which projected 1,350,000 square metres gross floor area for retail, housing, private and public office space (see Figure 13.1). Around 50 per cent of the total area has been finished to date (Belval 2018). Upon completion, Belval should host up to 25,000 employees, 7,500 students and researchers and 8,000 inhabitants (Agora 2018).

Today, the Cité des Sciences accommodates various institutions of higher education, public research, professional training and business innovation, in addition to public administration and about 100 private businesses (see Table 13.2). About 6,000 students and 5,000 employees are currently working on the site. Since 2013, two of the three faculties of the University of Luxembourg (founded in 2003) as well as three research centres, two public research institutes and several specialised training and business support organisations have moved to Belval. The Cité des Sciences adjoins an unfinished business district containing a mix of private residential and commercial areas, a vast park with a secondary school and two residential areas with a mix of single-family homes as well as privately owned and rental apartments to the west (see Figure 13.1). The latter lie on the territory of the municipality of Sanem with approximately 15,000 inhabitants.

The enlargement of Esch-sur-Alzette, however, is not limited to Belval. As late as 2015, the state announced the planning for the reconversion of a second former industrial site covering another 54 hectares to the east of Esch-sur-Alzette (Agora 2017). Thus, a fragmented polycentric city structure tightly interlinked with the adjacent municipalities of Sanem and Schifflange will eventually evolve in Esch-sur-Alzette, establishing the core of one out of three national growth poles.

FIGURE 13.1 Initial master plan for Esch-Belval

Source: Agora, copyright Buro Lubbers (2003)

TABLE 13.2 Knowledge hub Cité des Sciences in Esch-sur-Alzette

Higher education	Scientific institutions and research centres	Training and development	Supporting innovation
University of Luxembourg (UL)	Luxembourg Institute of Socio-Economic Research (LISER) Luxembourg Institute of Science and Technology (LIST) Luxembourg Centre for Contemporary and Digital History (C²DH) Luxembourg Centre for System Biomedicine (LCSB) Interdisciplinary Centre for Security, Reliability and Trust (SnT)	Institut Universitaire International Luxembourg (IUIL) De Widong Centre for professional development in the health sectors	Luxinnovation Fond national de la Recherche Incubator 'Technoport'

Source: Agora (2018)

The urban sustainability challenge: Capitalising on an abandoned brownfield to create jobs and to boost Luxembourg's knowledge economy

The development of Belval rests on the combination of three distinct, yet simultaneously occurring policy processes: 1) the reduction of development pressures related to the service sector in Luxembourg City and its periphery which, for their part, are coupled with the country's dynamic population growth and the substantial rise of daily commuters from across the borders; 2) the growing need for diversifying Luxembourg's economy against the backdrop of global economic changes; and 3) increasing levels of support for the establishment of the country's own tertiary education in the form of the University of Luxembourg. Thus the Cité des Sciences is not only the material manifestation of the Grand Duchy's structural shift towards the knowledge economy driven by innovation and research. Due to the predominance of economic motivations, the project offsets sustainability's multi-dimensional characteristics. Therefore, the project may face particular challenges to its sustainability ambition, not least given that it is hardly possible to achieve the different goals simultaneously (Dale et al. 2012; Flint and Raco 2012).

Locating the Cité des Sciences in the historic cradle of Luxembourg's industry has a strong symbolic meaning. Post-industrial economic growth and prosperity should no longer be generated through ore mining and steel production, but through data mining and the development of digital technologies.

This is expected to occur partly because of direct, place-based impacts that originate from the considerable public and private investments made in research, research facilities and campuses, and partly because of indirect or growth-related impacts such as networks, immigration or venture capital flows (Hall 1997). Following the rationale of urban anchor institutions, policy-makers posited that the materialised knowledge economy could simultaneously serve as a driver for urban and regional development in its various dimensions. The Cité des Sciences project consequently aims to achieve economic and partly social development goals determined almost exclusively at national level (Charrel 2014; Becker and Hesse 2013).

The sustainability challenge of Belval is to create an entirely new science district on a brownfield site next to a steel plant that is still operating, within a brand-new neighbourhood comprising cultural facilities, shopping malls and housing and in a context of old-industrialised patterns of urbanisation and settlement (Figure 13.2). It is undisputed that the development of a new district for the knowledge economy can provide a massive injection of social and economic life to a previously disadvantaged neighbourhood. The crucial question is probably how the 'new' site can be tied effectively to existing life worlds, economic systems and the existing built environment. Today, some 15 years following the start of the construction works in the Cité des Sciences, the interim results are mixed and not entirely satisfying. Whereas the quality of architecture and

FIGURE 13.2 Esch–Belval: Creating a knowledge city

Source: Le Fonds Belval (2018)

buildings is judged to be relatively high and spin-off and spill-over effects are slowly unfolding (Luxinnovation 2017; Technoport 2018), the site remains insufficiently integrated into the urban fabric of Esch-sur-Alzette. In the north, major traffic infrastructures and an existing residential area border the quarter. On the southern edge, the Franco-Luxembourgian border defines the area. To the east, the Cité des Sciences is bounded by a large electric steel plant that isolates it from the core inner city area. Only the northwest fringe of the area, where the park and the new housing districts have been developed, harbours some potential for a more organically grown addition of new land parcels to the existing settlements. However, overall interlinkages with the neighbouring municipalities, whether in terms of hard infrastructure or soft networks, remain underdeveloped.

Another key problem is the lack of non-work-related or non-consumerist venues and meeting places with good levels of recreational qualities for the employees, students, inhabitants and visitors of the Cité des Sciences. Belval is as yet merely an office town with restaurants that cater to the lunchtime demand of the white-collar workforce, rather than a mixed-use environment that provides space for everybody and brings life to the site in the evenings or on weekends. The building blocks that ought to speak to each other in such a mixed-use environment are rather big. Consequently, they hardly allow for a fine-grain, local synthesis of opportunities and activities to evolve. This problem stems largely from difficulties in connection with the financial and economic viability of the project. Despite the substantial investment assumed by the Luxembourg state for realising the Cité des Sciences, each plot of land beyond the latter had to be valorised under the conditions of the (private) real-estate market. On the supply side, major costs were incurred for cleaning up the former steelworks site and likewise for other relevant infrastructure. Refinancing was secured through successful development. On the demand side, land is scarce in Luxembourg, and recent economic and population growth have triggered a strong demand for office and housing space, hence turning the office market into the strongest competitor. The dense and compact development prepared under the rules of the real estate market therefore appears problematic.

The project is particularly challenging because the concept for the Cité des Sciences (including the locational choice) emerged primarily from state-induced decision-making. Planning decisions were implemented from above and barely coordinated locally (Leick 2016). Due to the lack of local participation (both in terms of municipalities and of civil society representatives), this posed a series of political difficulties, most notably in terms of representation. Planning was also pre-determined by a peculiar governance construction. First and foremost, the location decision for placing the university and other major research institutions in the south of the country was made in accordance with the then dominant vision of a more decentralised development pattern of the country (the so-called Integrated Concept for Regional Planning and Transport (IVL), introduced in 2004).

According to this concept of decentralised concentration, the south of the country with Esch-sur-Alzette and Belval at its core and the *Nordstad* (prospective) in the North were supposed to take growth pressure away from the crowded areas of the capital, Luxembourg City, and to provide a more balanced spatial pattern in the Grand Duchy. Second, at the local level, decentralisation had to be accompanied by a certain degree of concentration. This was justified with the need to realise a critical extent of agglomeration, in order to create a 'liveable' district. While overall decentralisation was not able to counteract the current centrifugal market mechanisms (i.e. demand for office space targeted predominantly towards Luxembourg City), the massive concentration of office space in Belval, a former site of heavy industry, stands in stark contrast to existing neighbourhoods and social worlds (Hesse 2013). Just like with similar initiatives such as inner-city brownfield or waterfront regeneration projects, it is difficult to link the formerly industrialised world with the service industry that has arisen at these sites. The majority of jobs created or relocated there can hardly meet the demands of the large local working-class population still present there and a fuller integration of these different worlds may take decades to materialise.

The planning innovation

Throughout history, the Grand Duchy has repeatedly revealed a high potential for innovation and strong problem-solving competences, especially with regard to the country's economic and political development. Luxembourg has displayed on several occasions a notable capacity to reinvent itself, most importantly by organising the transition from the steel and mining industries to becoming a major seat of European institutions and a financial centre of global importance. Alternatively, the quick pace of recent economic and demographic growth calls into question any serious attempt to become more sustaining and more sustainable. It also leads to structural challenges that are difficult to resolve. The processes within which Belval was conceived and ultimately implemented reflect the general challenges in terms of urbanism and spatial development that the country is facing. Nevertheless, depending on the definition of the term, the case also mirrors Luxembourg's capacity for planning innovation.

The editors of this book refer to Meijers and Thaens' definition of urban innovation as 'innovative practices within urban environments with the aim of improving those environments' (Meijer and Thaens 2018, 365). According to the Oxford Dictionary to 'innovate' means to 'make changes in something established, especially by introducing new methods, ideas, or products' and an innovative person is 'introducing new ideas; [he/she is] original and creative in thinking' (Oxford Dictionary 2018). These different definitions illustrate the ambiguity and emphasise the positive connotation of the concept of innovation. The ambiguity results from the relativity of the notion of 'new' and the blending of 'new' and 'original'. Introducing a new idea, for instance the idea of a mixed-use district, into a legislative zone planning setting, does not necessarily mean that the idea is new or original in general. To make changes by introducing a planning practice, which exists in a particular urban environment and a particular planning context, does not necessarily imply that the practice is inventive per se. Nor does it mean that the practice is adapted to the specific conditions of the new environment with the result that, basically, a new method is created (and new expectations are formulated). Innovation is implicitly linked to positive changes. But new methods and innovative ideas do not necessarily result in improvements or, as research on cases of urban innovation suggest, to (more) sustainable urban developments. The remaining part of this subsection will therefore explore the ambiguity of planning innovation by means of the analysis of the regeneration processes identified in the case of Belval in general and the Cité des Sciences in particular.

Innovation process

Policy-makers not only presented Belval and the Cité des Sciences as a place offering the best conditions for innovative activities; the planning process and the related governance structures that such a move towards innovation required were themselves declared to be forward-looking and highly innovative. In the parliamentary debate about the involvement of the state in the development agency to be created for the redevelopment of brownfield sites (Agora), the mayor of Esch-sur-Alzette and MP for the Luxembourg Socialist Workers' Party (LSAP) stated:

> We as the LSAP parliamentary group support the commitment of the government to contribute to herald the structural change in the South and we want to contribute that this first big test, the Belval brownfield, will become a model project for Luxemburg.
>
> *(L. Mutsch, former mayor, in Chambre des Députés 2001, 2650)*

This was, however, not the first time that the Luxembourg state initiated the development of a whole new city district. In the early 1960s, in an attempt to promote Luxembourg's status as a European capital, a public institution was created and commissioned to develop the 365 hectares on Kirchberg Plateau adjacent to the core area of Luxembourg City (Hesse 2013). In the context of the redevelopment of Belval, the actors involved dismissed the idea that important lessons could be drawn from the Kirchberg case (see for example Garcia and Mutsch in Chambre des Députés 2002, 2367–2370). At most, Kirchberg could serve as a negative example of a large-scale urban development project. The reasons for this reflection were manifold: first, the process was characterised by ad hoc decisions and a lack of planning in general. Second, Kirchberg has been developed in the spirit of modernist urban planning with land-use zones and a motor car-friendly infrastructure.

There was hence a general consent that the redevelopment of Belval should be approached in a completely new way.

The Belval planning process can be divided into different phases, each with varying actor networks and approaches. In 1999, after the election of a new government, competences around brownfield redevelopment were transferred to the Ministry of the Interior. Those involved in the project claimed that, in contrast to the previous phase (1997–1999), the first years following the change of competence were marked by important collaborative efforts between the representatives of the Ministry of the Interior, the municipalities and a local urban design office (Chambre des Députés 2002, 2668–2669). Simultaneously, the idea of creating a business park combined with zones for housing, leisure and culture in Belval was abandoned in favour of a sustainable urban neighbourhood with dense and compact building structures, mixed uses and short distances. Several large European urban development projects such as Rummelsburger Bucht in Berlin served as an example. It is during this phase that the intention to break new ground was most apparent.

The Ministry of the Interior negotiated behind closed doors also the framework for a public-private partnership (Agora) with ArcelorMittal, the owner of the steel company. At the time, this constituted a novelty in Luxembourg. By establishing Agora in 2000 and by adopting a law which allowed the Luxembourg state to acquire EUR €50 million worth of shares of Agora (i.e. 50 per cent of the shares), the governance structures were permanently changed. The planning approach had to be adapted to these changing prerequisites. Since one of the main objectives was to guarantee the financial viability of the operation, the management board's focus and understanding of a successful project largely differed from the municipal delegates' concerns. External experts from Berlin were appointed to establish an innovative financing scheme. A so-called cash flow model method was eventually introduced, which indicates that the reconversion of Belval was approached as a business project requiring clear-cut project management tools to steer the process. Upstream studies from local planners were used to test the feasibility of the project and financial calculations made in order to determine how many usable square metres of office, housing and commercial space had to be built and sold in order to cover the development costs. The Belval construction programme was established based on these financial calculations, not on social needs or on the wish to create a new district that would fit into the existing urban fabric. An international urban design competition was only launched afterwards. By making the indicators for the construction programme compulsory, the development company did not give much planning freedom to the participants. The competition was ultimately won by a multi-disciplinary team in cooperation with the Dutch planning office of Jo Coenen (Agora 2002).

This project management approach is actually more consistent with the mindset and action patterns of the private partner ArcelorMittal and, to some extent, of local political decision-makers, than with a communicative understanding of planning that includes iterative practices. This has considerable ramifications for the management of uncertainty and the open-mindedness to change (Leick and Müller 2018). In the case of Belval, it led to a lack of flexibility and experimentation during the implementation of the project. Insofar, Belval indeed represents a case of innovation, as a new planning method was introduced – implementing a large-scale urban project in order to conform to the planning ideal of the dense, compact European city. However, the planning process as such can hardly be considered to be innovative, given its rather mechanical and top-down orchestration, which created an isolated site that is difficult to connect to the existing neighbourhoods. A key lesson to be learned from this case is not to undertake big, ambitious projects when planning capacities and planning flexibility are missing. According to Van Heur (2010, 1718), innovation in the context of higher education and research is increasingly understood 'as a complex and interactive process with many feedback loops, between multiple actors and organisations'. Conversely, the chances for innovation in urban planning processes are lower when actors try to stick to a linear planning approach as much as possible.

Trade-offs

A couple of sustainability trade-offs evolved during the planning and implementation of the Cité des Sciences and the entire site of Belval. These are somehow difficult to detect, given the hybrid character of the whole project as a flagship development of the state on the one hand, which must follow by and large the imperatives of the real estate market on the other hand. While an active and open management of conflicts did not exist, a more improvisation-based attitude to introducing necessary adjustments of the master plan was adopted in response to changing framework conditions. A first important trade-off appeared with regard to time and the related phasing of the development, which was designed at large scale and thus prone to uncertainty (Karadimitriou et al. 2013). The uncertainty can become critical if there is a certain lack of flexibility in the plan or the planning process. Development goals were quite ambitious, and, in fact, it is stunning that the new district including parts of the university were inaugurated less than 20 years after the closing of the steel mill and the start of the preparation works for the regeneration of the site. Even when disregarding the various delays that had occurred before the university's final move to Belval in summer 2015, such pace of implementation is impressive.

However, the pressure to capitalise on public and private investments, the ever-shifting nature of real estate cycles and the ramifications of the financial crisis of 2008/2009 were limiting the achievement of further sustainability goals. While the planning for and the realisation of the buildings of the Cité des Sciences, mostly coordinated by the public Fonds Belval and other government institutions, followed its own logic of implementation, the young university was confronted with plans prepared and implemented mostly from above. Thus, the envisaged time for realisation did not effectively allow the site to grow slowly, or users to become acquainted to the site step by step. Neither did it offer some 'free space' (for instance for use by students) whose definitive destination could be decided later on. The most critical issue was probably the instalment of two different development entities, the state-owned Fonds Belval for the Cité des Sciences on the one hand, and Agora for the development of the whole site. The transaction costs to be paid for two different bodies in charge of an overlapping territory were effectively high. The desired clarity in decision-making under difficult market conditions also led to a sort of rigid policy practice to emerge, which seems quite common in the small country. This happened at the cost of the active participation of the two municipalities and of the interested public.

What started to be a fully 'integrated' plan for an urban quarter became more of a selective endeavour, within which market forces and concuring interests became more or less predominant for its outcome. This points particularly at the relationship between Fonds Belval and the University as its main customer on the one hand, and between Agora and commercial investors on the other hand. Consequently, social sustainability became under pressure due to the need to bring major parcels of land to development, mostly under market conditions. Since the whole project was initially broken down into a few major building blocks, the changing demand for smaller, more flexible units (e.g. for office use) could not be met. (The latter approach is foreseen in future development cycles). For a certain period of time, high-end housing units were placed on the market more pro-actively, which has proven to be successful until recently. However, only a bare minimum of social housing could be realised, with the rest of the housing being designed to attract the localisation of an 'intellectual middle class' with a certain purchasing power, regardless of whether the urban amenities this target group is looking for can be provided on the site or not. Future plans of the state-owned social housing agency SNHBM for a development in Belval-Nord (64 units) have been announced to start some time in 2019.

While decentralised concentration was used as a guideline for the spatial policy that led the government's national planning approach (IVL), compact city principles should bring life to the site and ensure a critical mass of employees, inhabitants and customers locally. The related scalar mismatch – or trade-off – is significant: what may have made sense from a national development perspective (rational use of land, brownfield development, supporting the transformation of the old industrialised South) is not necessarily appropriate at the fine-grain yardstick of local development (also given that the site was considered by the City of Esch-sur-Alzette to be only second best for the new development).

High-density development also leads to congestion and thus trade-offs against the perceived benefits of concentration. For this reason, a new train station, 'Université-Belval', and a huge park-and-ride facility were provided. At the same time, the motorway network was expanded, in order to improve the transport links between Belval, France (Liaison Micheville) and Luxembourg City (A4-Motorway). In this light, it seems rather likely that the predominantly car-based transport behaviour of the Grand Duchy's residents and of the large majority of foreign commuters will remain unchanged; across the European Union, Luxembourg has the highest motorisation rate (Eurostat 2018). Bike infrastructure exists as well, e.g. parking spots for share-bikes, and bike routes or paths in Belval are included in official promotion materials. However, the overall connectivity and accessibility of the site by bike is of rather poor quality and has not improved recently. This is a result of both the existing steel mill functioning as a physical barrier between Belval and the urban centre of Esch, and also of the predominance of motor traffic in the country's road design standards and practices. These issues are hard to change and may alter only in the medium or long term.

Significant sustainability trade-offs are also visible in terms of its ecological dimension. High-density development placed in a setting that is designed to adapt compact city principles can easily materialise at the cost of environmental quality. While formal assessments of the built environment claim to meet highest sustainability standards (Belval 2011; Le Fonds Belval 2011a), there are some nuisances at the micro-scale. High buildings tend to create wind turbulence that lowers the quality of the use of public space – which in design terms was actually a prominent target of the development bodies. Green spaces as such are limited in the Cité des Sciences. Full green space compensation is provided by Park Belval. However, the park is situated almost a kilometre away from the core site, and its stimulating effect may not come to fruition for those who work, teach and study in the Cité des Sciences.

Probably the most important and most challenging trade-off between the economic, social and environmental dimensions of sustainable urban development practice is the convergence of the different life-worlds of the former steel workers (many of whom still live in the surrounding municipalities) on the one hand, and the new services class that has arrived to (at least) work in Belval on the other hand. It is also illustrated by empirical evidence that the former industrial sectors hardly profit from such injections, even though there might be multiplier effects that result from the development of retail and office sectors that can also offer potential for new jobs. It remains to be seen in the longer term whether, and if so, how far, a transition is possible that provides benefits to both social worlds that are prevalent here, i.e. the blue-collar and the white-collar workers. It is difficult to argue whether such broader social trade-offs or conflicts can be resolved by developing certain mechanisms of management and communication. In the young history of Belval, these are as yet missing.

Conclusion

To conclude, all chapters in this book are called upon to discuss whether the planning innovations that have been identified in their respective case descriptions are transferable to other contexts and whether the findings result (or not) from factors that could be replicated elsewhere. The scrutiny of the planning innovations inherent in the case of the Cité des Sciences is not an easy matter. Given the complexity of such a development, we tend to be cautious in terms of causal dependencies, linear relationships and direct transferability. So, the answer to the aforementioned question is actually rather no than yes, since, as is shown by the case of Luxembourg, the factors leading to the particular outcome cannot be replicated easily. The country's political and socio-economic specificities in addition to its very distinctive development path need to be taken into account when assessing the Belval project. Whereas Luxembourg is a small but highly globalised place of service and finance industries that has experienced stunning growth rates over the last decades, the planning system as such, established only recently, lacks a sufficient track record for Belval to be treated as a template case.

When assessing the project in overall terms, one needs to acknowledge that what has been created on the Belval site so far looks, on the one hand, really impressive. Within just 15 years, a highly modern and future-oriented knowledge hub interspersed with housing, retail and cultural facilities has been built on an almost empty 'canvas'. This is confirmed by the growing number of visitors and scholars from Luxembourg and abroad who come and marvel at the ensemble of architecture and industrial heritage displayed here. With the site getting more complete in the future, some of the issues mentioned above might be resolved. However, if we look beyond the mere physical manifestations of planning innovation in Belval, the glass appears half-empty. Especially the Cité des Sciences, the district's core area, lacks vibrancy, amenity values and social diversification. To be fair, social processes of appropriation cannot be created artificially but develop incrementally over time. The reasons for this deficit are rather of a structural and conceptual nature, given the project's high ambition, the usual risks that arise from large-scale urban development, and the overarching economic interests that have driven the project. Such contradictions are not limited to the case presented here (see Chapter 1), but unfold in Belval quite distinctively.

Finally, the question is what could be learned from the case of Belval in terms of the potential transferability of the project's outcome and experience? This seems less relevant an issue for other places abroad. Learning would primarily make sense at a national level, most notably as to the critical role of large-scale projects pursued in a small country, and also with particular respect to the planned reconversion of the old-industrialised site in Schifflange, the municipality to the southwest of Esch-sur-Alzette. Here, the municipalities and developers in charge, who are to a large extent identical to the stakeholders that were concerned with the Belval project,

194 Becker, Hesse and Leick

have the chance to demonstrate what learning from Belval could mean: to aim for less rather than more; to design a more adaptive and flexible project; and to practice a more inclusive and participative planning process. This could help to create liveable and thus sustainable urban neighbourhoods, which are deemed essential for the country's future (cf. Rydin 2010).

References

Agora (2018) Belval Market Figures 2018/1 (www.agora.lu/fr/infotheque/) accessed 19 April 2018
Agora (2002) "Concours pour la réalisation de l'urbanisation de Belval-Ouest Esch-sur-Alzette/Sanem (Belvaux) Luxembourg / Städtebaulicher Realisierungswettbewerb Belval-Ouest Esch-sur-Alzette/Sanem (Belvaux) Luxembourg", *Documentation et perspectives /Dokumentation und Perspektiven*, Agora
Agora (2017) "Projet de reconversion de l'ancien site sidérurgique d'Esch-Schifflange - Rapport de synthèse des études environnementales et des diagnostics préalables", *Rapport de synthèse des diagnostique Site Esch-Schifflange*, Agora
Becker T. and Hesse M. (2013) "Building a sustainable university from scratch: anticipating the urban, regional and planning dimension of the 'Cité des Sciences Belval' in Esch-sur-Alzette and Sanem, Luxembourg" in König A. ed., *Regenerative Sustainable Development of Universities and Cities: The Role of Living Laboratories*, Edward Elgar, Cheltenham, 254–272
Belval (2011) "Quartier labellisé" (www.belval.lu/fr/belval/quartier-labellise/) accessed 25 April 2018
Belval (2018) "Le nouveau Belval" (www.belval.lu/fr/belval/le-nouveau-belval/) accessed 19 April 2018
Carrière J. P. and Demazière C. (2002) "Urban planning and flagship development projects: lessons from Expo 98, Lisbon", *Planning, Practice & Research*, 17, 1, 69–79
CEFIS (2013) "Fiche commune 'Esch-sur-Alzette'" (www.cefis.lu/resources/15-Esch-sur-Alzette.pdf) accessed 19 April 2018
Chambre des Deputés du Grand-Duché du Luxembourg (2001) "Compte-rendus des séances publiques de la Chambre des Députés", *69e séance*, Chambre des Deputés du Grand-Duché du Luxembourg
Chambre des Deputés du Grand-Duché du Luxembourg (2002) "Compte-rendus des séances publiques de la Chambre des Députés", *64e séance*, Chambre des Deputés du Grand-Duché du Luxembourg
Charrel M. (2014) "Face à la fin du secret bancaire, le Luxembourg cherche à diversifier son économie" (www.lemonde.fr/economie/article/2014/10/16/face-a-la-fin-du-secret-bancaire-le-luxembourg-cherche-a-diversifier-son-economie_4507345_3234.html) accessed 19 April 2018
Dale A., Dushenko W. T. and Robinson P. (eds) (2012) *Urban Sustainability. Reconnecting Space and Place*, University of Toronto Press, Toronto
European Commission (2013) "Urban development in the EU: 50 projects supported by the European Regional Development Fund 2007–2013" (http://ec.europa.eu/regional_policy/en/information/publications/studies/2013/urban-development-in-the-eu-50-projects-supported-by-the-european-regional-development-fund-during-the-2007–13-period) accessed on 19 April 2018

Eurostat (2018) "Passenger cars in the EU: statistics explained" (http://ec.europa.
eu/eurostat/statistics-explained/index.php/Passenger_cars_in_the_EU) accessed 27
April 2018

Flint J. and Raco M. (2012) *The Future of Sustainable Cities: Critical Reflections*, The Polity
Press, Bristol

Gold J. R. and Gold M. M. (2008) "Olympic cities: regeneration, city rebranding and
changing urban agendas", *Geography Compass*, 2, 1, 300–318

Grin J., Rotmans J., Schot J. in collaboration with Geels F. and Loorbach D. (2010) *Transitions
to Sustainable Development: New Directions in the Study of Long-Term Transformative Change*,
Routledge, London

Hall P. (1997) "The university and the city", *GeoJournal*, 41, 4, 301–309

Hesse M. (2013) "Das 'Kirchberg-Syndrom': grosse Projekte im kleinen Land: Bauen und
Planen in Luxemburg", *disP-The Planning Review*, 49, 1, 14–28

Jones P. and Evans J. (2013) *Urban Regeneration in the UK: Boom, Bust and Recovery*, Sage,
London

Karadimitriou N., de Magalhães C. and Verhage R. (2013) *Planning, Risk and Property
Development: Urban regeneration in England, France and the Netherlands*, Routledge,
London

Le Fonds Belval (2011a) "La Cité des Sciences sous la loupe", *Magazine* 3

Le Fonds Belval (2011b) "La cité des sciences - un projet en évolution", *Magazine* 2

Leick A. (2016) "Kleines Land, große Projekte. Diskurse, Praktiken und soziale Welten
im Entscheidungs- und Planungsprozess der Großvorhaben Belval und Kirchberg in
Luxemburg", Unpublished PhD thesis Department of Geography and Spatial Planning,
University of Luxembourg

Leick A. and Müller M. (2018) "The problem with projects: (not) thinking urban development through projects", unpublished working paper

Luxinnovation (2017) "Annual Report 2016" (https://www.luxinnovation.lu/global/2536/)
accessed 19 April 2018

Meadowcroft J., Farrell K. N. and Spangenberg J. (2005) "Developing a framework for
sustainability governance in the European Union", *International Journal of Sustainable
Development*, 8, 1/2, 3–11

Meijer A. and Thaens M. (2018) "Urban technological innovation: developing and testing
a sociotechnical framework for studying smart city projects", *Urban Affairs Review*, 54,
2, 363–387

Moulaert F., Salin E. and Wequin T. (2001) "Euralille: large-scale urban development and
social polarization", *European Urban and Regional Studies*, 8, 2, 145–160

Oxford Dictionary (2018) Definition "innovative" (https://en.oxforddictionaries.com/
definition/innovative) accessed 13 April 2018

Ponzini D. (2011) "Large-scale development projects and star architecture in the absence of
democratic politics: the case of Abu Dhabi, UAE", *Cities*, 28, 3, 251–259

ProSud (2017) "Démographie Sud" (www.prosud.lu/documents/download/317/indices-sud-
4---d%C3%89mographie) accessed 19 April 2018

Rydin Y. (2010) *Governing for Sustainable Urban Development*, Earthscan, London

Statec (2017) "Les communes de la région sud", *Recensement de la population* 26, Statec

Statec (2018) "Etat de la population" (www.statistiques.public.lu/stat/ReportFolders/Report
Folder.aspx?IF_Language=fra&MainTheme=2&FldrName=1) accessed 19 April 2018

Swyngedouw E., Moulaert F. and Rodriguez A. (2002) "Neoliberal urbanisation in Europe:
large-scale urban development projects and the new urban policy", *Antipode*, 34, 3,
542–577

Taşan-Kok T. (2009) "Entrepreneurial governance challenges of large-scale property-led urban regeneration projects", *Tijdschrift voor Ekonomische en Sociale Geografie*, 101, 2, 126–149

Technoport (2018) "Incubator, co-working, fab lab" (www.technoport.lu/online/www/function/homepage/ENG/index.html) accessed 24 April 2018

Thewes G. (2017) "About . . . the history of Luxembourg" (www.luxembourg.public.lu/en/le-grand-duche-se-presente/histoire/histoire-mots/economie-1945/index.html) accessed 24 May 2018

Van Heur B. (2010) "The built environment of higher education and research: architecture and the expectation of innovation", *Geography Compass*, 4, 12, 1713–1724

14

CONCLUSION

Sébastien Darchen and Glen Searle

Sustainability and urban planning practice

As stated in the Introduction, the 12 case studies presented in the book provide responses to 8 of the 17 sustainability goals stated in the 2030 Agenda for Sustainable Development presented by the United Nations. This does not mean that the planning innovations analysed in this edited volume are the only answers to the sustainability challenges presented in the 2030 agenda. The planning innovations analysed should be considered as avenues for change but they are the not the only solution to a given sustainability challenge. In that sense, the book highlights the prevalence of contextual factors in shaping planning innovation in a given context. One of main research questions that this edited volume aimed at answering was: *Has the concept of sustainability made a difference in urban planning practice?*

The book demonstrates that the concept of sustainability, even if some would argue that it is still a contested concept for the development industry in the context of brownfield regeneration for example (Dixon 2006, 260), has indeed made a difference in urban planning practice. The book shows that some innovations such as the one analysed in Chapter 2 are still at the stage of a conceptual innovation. Mechanisms to ensure that the innovation (policy co-production and Indigenous development) can be implemented are still not in place. As analysed in Chapter 2, the reconciliation of cultural differences in urban development is still hindered by colonial beliefs from community planners and politicians. By contrast, some planning innovations have had a long life and stood the test of time, as illustrated in Chapter 7 on the waste management strategy in the Indonesian context that is still implemented today after 13 years. On this point, planning innovations either driven by the community (Chapter 7) or with significant input from the citizens (Chapters 3, 4, 5, 6, 7, 11 and 12) are more resilient and better 'designed' to respond to a specific sustainability challenge. Innovation processes that did not integrate citizens adequately in the re-use of infrastructures (restoration project in Seoul and adaptive re-use in Los Angeles) or in the creation of a new infrastructure (Science City in Belval, Luxembourg) (Chapters 9, 10 and 13) have had shortcomings in regard to social sustainability: destruction of social networks due to relocation (Seoul) and gentrification (Los Angeles), and a lack of vibrancy (Science City in Belval). Overall, the case studies presented demonstrate that sustainability as a concept is still very much alive and relevant to planning practice.

Transitions to sustainability

The chapters presented in the book highlight the importance of contextual factors in enabling planning innovations to happen. A main research question that this book aimed at answering was: *What elements of the city's political ecology and economy, in terms of power relations, institutional settings and so on, assisted or were in conflict with the transition agenda and, in particular, the sustainability innovation?*

Contextual factors

Contextual factors refer to the city's political ecology and economy, in terms of power relations and institutional settings that are facilitating (or inhibiting) the transition agenda to happen. This transition agenda can have a relatively long timeframe, as is explained in the case of Vancouver (Chapter 5) with the development of a more inclusive process for planning and policy co-creation. The innovation unfolded over 20 years and the Vancouver model of local government has shown some flexibility in enabling the planning innovation to develop and improve over time. The positive influence of contextual factors in the development of the transition agenda is even more obvious in the Freiburg case study (Chapter 6). Sustainability was appropriately positioned in the municipal administration; in addition, a concession levy appropriation of 10 per cent was created for sustainability policy development. Some sustainability initiatives, as shown in Chapter 9, have a dimension that goes beyond the city itself. The Cheonggyecheon Restoration Project in Seoul had a national dimension to it because of its symbolic significance as the restoration of a major infrastructure in the city. Chapter 11 (Bhubaneswar) also highlights the municipal scale in shaping an innovative approach for the implementation of the Smart City scheme sponsored by the national government. It is also the case in the Belval project that was initiated by the Luxembourg state (Chapter 13). External experts were also appointed from Berlin to establish an innovative financing scheme. However, in most of the planning innovations analysed the contextual factors are more prevalent than the external factors in explaining the emergence and refinement of innovations. Most of the planning innovations evolved over time and most of them have a long timeframe, over 10 years (Chapter 5 Vancouver; Chapter 6 Freiburg; Chapter 7 Indonesia; Chapter 9 Seoul; Chapter 10 Los Angeles).

Some of the planning innovations presented were enabled through funding opportunities created to facilitate the implementation of the innovation (Chapters 5, 6 and 10). In the case of Vancouver, the Community Economic Development Strategy – a new economic framework informed by low-income residents – was resourced through a self-organised co-creation committee with CAD$150,000 over three years. This represents an innovation, as before the resources were used by the City to share research, inform and consult; instead, the choice was made to use the resources for residents to do their own research. The authors stress the shift from externally oriented efforts in firm and talent attraction to more attention being placed on local exchange with low-income residents and more resources for residents to develop their own process for engagement and deliberation (Chapter 5). In the same vein, as mentioned above in the case of Freiburg (Chapter 6), a concession levy appropriation of 10 per cent was created to ensure a long-term and successful sustainability policy. In the case of waste management in Yogyakarta, the initiative required external funding to start up. A private donor from Australia provided a small amount for the project to start but over time the community demonstrated leadership and owned the innovation, which also indicates that contextual factors are key in shaping innovations. In the case of the adaptive re-use strategy in Los Angeles, financial incentives helped pioneers of adaptive re-use like Tom Gilmore: his first adaptive re-use project benefited from public funding from the City. Authors for the Belval project (Chapter 13) also confirmed that to help with finance, external experts were appointed as a new cash-flow model method was required.

Most of the planning innovations presented were city-led (Chapters, 2, 3, 5, 6, 9, 11 and 12). Some planning innovations like the Belval project (Chapter 13) have a European dimension so the actors' networks mobilised go beyond the city. This is also the case with the restoration project in Seoul (Chapter 9) and with the Freiburg experience (Chapter 6). One innovation is community driven – waste management in Yogyakarta (Chapter 7). Some cases are hybrid with a strong involvement of the residents but still city-led, as in the cases of Helsinki (Chapter 3), Vancouver (Chapter 5), India (Chapter 11) and San Francisco (Chapter 12). The only example strongly relying on the private sector (developers) is the adaptive re-use strategy in Los Angeles (Chapter 10).

External factors

This edited volume demonstrates little impact from external factors in the development of planning innovations. Rather, planning innovations are the product of an endogenous process. The findings of the book align with the literature on sustainability transitions that stresses the centrality of context in determining the evolution of niche innovations (Loorbach and Rotmans 2010; Rutherford and Coutard 2014).

In this volume, we make a distinction between contextual and external factors. External factors can be associated with the concept of *mobile urbanism* and the transfer of sustainability policies from one place to another. External factors also refer to the national and international networks in which cities are involved and that might trigger the development of planning innovations locally. Paradoxically, in a global world, planning innovations are mostly the product of local factors: Chapter 2 highlights the role of the municipality in transforming a conceptual innovation into an implementable solution; in Chapter 3, the authors highlight an increased autonomy of the city of Helsinki (through the Finnish Land Use and Planning Act, 2000) that played a role in the development of the planning innovation. In Chapter 4, the author acknowledges the role of national security policies in enabling all aspects of human security, but social urbanism as an innovation has been shaped in collaboration with the community and local actors. Sotomayor explains that Proyecto Urbano Integrado (PUI) is an area-based planning instrument that was inspired by the Barcelona model of urban regeneration and Bogota's public space and citizenship culture programs from the 1990s. Overall, it is the participation of local actors that has shaped social urbanism as an innovation in Medellin. In Chapter 5, the authors show that making resources available to level the balance of power in planning processes was the innovation for the Vancouver Community Economic Development (CED) strategy. This strategy empowered local actors and launched the innovation.

The mobilisation of civil society has been identified as an important factor for the emergence of innovations, as was the case in Freiburg (Chapter 6) but also in Seville (Chapter 8). Even in the case of waste management in Indonesia where external funding (a small amount) was necessary to launch the innovation, the long life of the innovation relies on the leadership of the community (Chapter 7). Three case studies: the Seoul restoration project (Chapter 9), the implementation of the Smart City concept (Chapter 11) and the Belval project (Chapter 13) describe initiatives initially either driven by or supported by the national government. However, the authors have highlighted the need to foster the involvement of the local actors in making these initiatives a success in the long run. For example, Chatterji and Maitra (Chapter 11) explain how the chances of success of a centrally sponsored scheme are enhanced if the planning process also reflects a participatory and inclusionary approach.

Trade-offs

Chapter 1 highlighted Campbell's (1996) conceptualisation of nascent conflicts between the three legs of sustainability leading to the identification of trade-offs between the three legs as a potential outcome of sustainability innovations. The case studies in this book suggest that it is possible for trade-offs to be avoided, but that certain preconditions might be needed if this is to happen. This section overviews the book's findings concerning trade-offs.

The innovation processes in which there is explicit incorporation of more than one dimension of sustainability are, most obviously, those where trade-offs are least likely. Several case studies in this book show that strong participatory policies that allow communities to have a central role in decision-making and thus contribute to social sustainability can also allow positive outcomes on other sustainability dimensions to be achieved. This incorporates the need for sufficient resources to be allocated to allow participation to achieve such a role. In Vancouver (Chapter 5), a tradition of collaborative participatory processes was matched with resources to bring about inclusive economic revitalisation that developed employment opportunities targeted at low-income groups. Thus both economic and social sustainability outcomes were achieved.

The Bhubaneswar case study (Chapter 11) illustrates a similar situation on a larger scale. The Smart City Plan envisages the transformation of the overall economy toward business services and tertiary education. But it also incorporates actions to improve the livelihoods of the poor through skill-building and support programs for micro-enterprise incubators and street vendors, plus improving basic services in slums and awarding property rights to slum dwellers. Land use policies are also promoting transit-oriented development. Thus, the Smart City Plan addresses each of the three sustainability legs, and uses a creative approach to participation to achieve this. The Plan and its various processes have been underpinned by a combination of favourable institutional factors including national government financing for Smart City planning, and a state and city government socio-political ideology that is conscious of the needs of the urban poor. Like Vancouver, this has allowed sufficient resources for socially sustainable participation, but also allowed even more comprehensive sustainability outcomes.

The sustainability innovation in Yogyakarta, Indonesia (Chapter 7) has achieved positive outcomes on all three dimensions of sustainability, with strong local participation again central to such outcomes. The project targeted waste management to improve environmental sustainability by empowering the local community through capacity building, which ultimately resulted in new livelihood opportunities. As in many of the other innovations in this book, the existence of local leaders who could act as agents of change (in this case, to provide environmental counselling and guidance on waste management) was essential to success.

A local tradition of strong support for sustainability initiatives can produce innovations that positively address all three dimensions of sustainability. In Freiburg (Chapter 6), trade-offs across all three sustainability dimensions were minimal in the development of a brownfields precinct. The city's support of sustainability from the early 1970s emerged from its well-educated public with many employees in education, science and research generally. This resulted in a municipal council focused on sustainability initiatives, supported by mayors and department directors who promoted projects and processes and motivated other stakeholders. These underlying factors have then been reflected in specific conditions that have produced innovative sustainability outcomes in brownfield redevelopment, such as strong minimum sustainability standards for development, and citizen participation as a central element of redevelopment processes.

The Belval Luxembourg project (Chapter 13) has similarities as a brownfield redevelopment initiative with the Freiburg case study, in this case involving the planning ideal of the dense, compact city plus a new knowledge-based economy. However, contextual factors favouring highly sustainable outcomes have not been as firmly in place as in Freiburg, and thus some trade-offs have emerged. Public participation has been limited, while the ambitious scale of development that was to be achieved relatively quickly caused a need to bring major parcels of land to development largely under market conditions, with resulting sustainability issues. Higher income housing has been favoured at the expense of social housing; the motorway network was expanded to improve access for in-commuters; most green space is at some distance from the core; and the project does not seem to benefit the former steel workers who used to have jobs on the site.

The Medellin case (Chapter 4) illustrates a context in which civil society has become so unsustainable that it is negatively affecting local economic sustainability. In such a situation, innovative processes to improve social sustainability also have a positive trade-off in increasing economic sustainability. Innovative policies to reduce security threats in marginalised Medellin communities not only improved employment outcomes in those communities, but were seen by the wider business community as an opportunity to lower their security costs and attract foreign investment.

A basic context in which trade-offs between the three legs of sustainability can emerge occurs when the innovative policy or practice results in the displacement of existing communities and economic activity – a situation not present in the above cases. This is illustrated by the Cheonggyecheon restoration in Seoul described in Chapter 9. There, the reinstatement of the natural waterway, with a variety of resulting environmental benefits, caused the rents of local merchants to increase and force their relocation. While some compensation was made available, full compensation would have made financing the restoration impossible. In addition, the increase in property prices caused gentrification and the destruction of social networks of pre-existing low-income residents. In this case, the trade-offs were typical of environmental improvement projects in major central city areas that induce large increases in property values valorising environmental improvements. In Cheonggyecheon, these trade-offs were generated despite a strong participation process involving numerous consultations that was backed by strong leadership and financial resources.

An analogous outcome occurred in the adaptive re-use innovation in inner Los Angeles (Chapter 10). There, the re-use ordinance applied to vacant office buildings, so that there were no discernable negative economic trade-offs for the environmental sustainability gains from shifting Los Angeles' sprawling development towards one that is denser, walkable and more sustainable. But the increase in property prices has led to gentrification and contributed to the removal of low-income residents, with negative social sustainability effects. Again, in similar fashion to the Seoul restoration project, this trade-off emerged under the weight of consequential increases in property prices despite local leadership of the process: in the Los Angeles case this included local non-governmental organisations (NGOs) helping to bring about the re-use ordinance itself. The San Francisco case study (Chapter 12) is an example where the sustainability focus on one dimension has rather limited trade-off implications for the other dimensions. The innovative policies to protect industrial space from office development and keep it for advanced and consumer-led manufacturing were intended to promote sustainable economic development by providing a broader range of employment beyond the city's booming high-tech and advanced services sectors. The policies emerged from a favourable local context that included a history of local advocacy, informed planners and the NGO SFMade. The latter has rebranded manufacturing as crucial to local identity and, as Grodach and Martin state in Chapter 12, "a viable source of living wage employment". In these terms, the initiative could also be argued to support the city's social sustainability. The Seville cycling mobility project (Chapter 8) is an even clearer example of trade-offs between sustainability dimensions being absent. Rather, trade-offs existed within the environmental dimension, with increased cycling mobility traded off against reduced motor vehicle priority.

Finally, in some cases trade-offs are so great that the innovation is precluded from being implemented. Most such cases probably involve innovations that threaten economic sustainability. In the Calgary case study in Chapter 3, the attempt to co-produce urban policy with local Aboriginal communities was essentially thwarted by the imperative of economic development, with the concerns of Indigenous elders about building on a historic tribal site being largely ignored and not given a formal role in the planning process. Here it was not so much a case of major economic gains being threatened, as a failure to change power relationships that ensured priority for economic development and did not give a formal voice to indigenous concerns.

Transferability

The enduring significance of the sustainable urban innovations described in this book lies largely in the extent to which they can be transferred to other contexts, and whether the outcomes result from factors that can be replicated elsewhere. In this section, the findings on this from the book's case studies are briefly summarised.

Several of the innovations in the book's case studies have already been applied in other urban areas. The public participation GIS that was applied in Helsinki has been used for planning information in many other countries, involving over 400,000 participants. It is a relatively simple tool that has enabled public participation to incorporate a greatly increased level of participation compared to older approaches, and produce versatile, high-quality data that can be used at different stages of the planning process. While public participation was the focus of the sustainability innovation in Helsinki, the significance of participation as a necessary condition for success and thus for transferability is a recurring theme in most of the other innovations in the book.

Another innovation that has already been widely implemented in modified form is the community-based waste management program in Yogyakarta, Indonesia. The innovation is technically fairly simple, but needs a context of community engagement and participation and of leaders who will act as 'agents of change'.

The Seville cycling mobility initiative required a comparable context, one that should be able to be replicated elsewhere. This context involved a strong local cycling association with a commitment to promote cycling in the city, and a supportive local government with a commitment to develop appropriate cycling infrastructure.

Some innovatory practices will require major institutional and political changes in order to be applied in new locations. Medellin's comprehensive urban policy approach to reduce violence and marginalisation has potential applicability to a range of cities in developing countries that face similar issues. However the necessary responses are daunting and involve significant shifts in urban governance practices incorporating much-enhanced public investment in services and infrastructure, institutional reforms in justice and policing, changes in state-community relations to build state legitimacy and authority at the local level, broad and sustained political will, and intervention anchored in community participation.

In the case of Freiburg, the factors promoting the city's successes in innovative sustainable development were also wide in scope, but evolved over a long period. They have generated what is essentially a sustainability ecosystem. This has a number of synergistic elements such as an active and educated community and associated scientific institutions resulting in strong municipal support for sustainable initiatives, sustainability leaders, significant regional government support and a high level of community participation, inter alia. This complex of favourable factors would need a number of years to be fully replicated in other cities, although some of these factors will already be in place in certain locations.

The Smart City project in Bhubaneswar, India, has a similarly broad sustainability scope as that in Freiburg. But it has not drawn on a similar long evolution of favourable factors although a social-democratic ideology prevailing at state and municipal level, involving a participatory and inclusionary approach, was a necessary starting point. The involvement of the national government, which instigated the Smart City scheme and provided 50 per cent of project funding, was critical. These factors are certainly distinctive, but variants of them could be present elsewhere.

For many of the innovations reported in this book, the necessary governance context for successful transfer is less demanding than comprehensive sustainability initiatives such as Freiburg's, in particular. The emulation of Vancouver's participatory community economic development strategy would require a similar local socio-political foundation of participatory planning that empowers marginalised groups, backed by the provision of adequate resources for use by such groups. In Cheonggyecheon, Seoul, extensive consultation and participation was central to the project's success, though in this case it emanated from strong mayoral leadership. The mayor drove the project forward, backed by strong financial resources and new institutional research capacity to demonstrate the worth of the project. The project was quite divisive, and an equivalent development elsewhere would therefore need the same strong leadership and political will, allied to extensive consultation, that allowed the Cheonggyecheon project to be completed.

Sustainability innovations also very often require certain economic contexts to allow their transfer to other cities. Cheonggyecheon renewal was possible partly because the Seoul city government was large enough to have the necessary financial resources to carry out the project. The success of the Los Angeles adaptive re-use ordinance was dependent on a city economy with an available stock of old buildings and a potential user population of young people committed to an urban setting and largely self-employed or working from home. Other cities where similar initiatives have been taken have a comparable economic context. San Francisco's industrial space preservation initiative is also the product of a particular economic context. The resurgence of manufacturing there has drawn on the area's entrepreneurial culture, design and engineering expertise, and a strong artistic and maker community. This has increased demand for industrial land in a wider context of surging tech company demand for office space. As in Los Angeles, NGOs (as well as community advocacy in San Francisco) have been vital to the development of innovative policies within the city's economic context. The applicable San Francisco context is more distinctive, however. The possibilities of replicating the manufacturing space preservation initiative elsewhere might therefore be reduced.

In some cases, the context for the innovation and the generative factors involved are so specific and distinctive that the prospects for transferability are very limited. This is the case for the Belval, Luxembourg project. There, the country's unusual political context and its singular economic development path, plus a relatively immature planning system, suggest that kind of innovation cannot be replicated easily. Even so, Becker, Hesse and Leick suggest in Chapter 13 that lessons from Belval could enlighten planning for a proposed similar brownfield project within the same country.

In other situations, difficulties in making the innovation happen because of insurmountable problems with trade-offs on other sustainability dimensions might also exist in similar contexts elsewhere. This seems exemplified in the case of the unsuccessful attempt for innovation that involved co-production of Calgary urban policy with local Aboriginal communities. Belanger, Dekruyf and Walker in Chapter 2 cite others to note that Calgary's situation is replicated in cities and towns across colonial countries including Australia and New Zealand, "whose leaders framed Indigeneity and urbanism as . . . incommensurable".

Concluding comments

In opposition to the literature that highlights the transferability of policies from one place to another (McCann and Ward, 2011), the book demonstrates that planning innovations are mostly the result of an endogenous process, often the result of a long tradition in a specific area of sustainability (Freiburg chapter) or the results of the mobilisation of civil society for the achievement of specific sustainability goals like the reduction of motorised traffic (Seville and Indonesian chapters, and also the chapter on San Francisco). In that sense the planning innovations are very much context specific.

Some chapters (Chapters 2, 5 and 12) have identified the benefits of co-creation with the end-users in effective urban policy-making for sustainability, even though in the example of Calgary the innovation is still at a conceptual stage. This is another indication that planning innovations mostly reflect the city's political ecology in terms of power relations and not the transfer of sustainability policies from one context to another.

The findings of the book contradict, to a certain extent, current statements on the high mobility of sustainable urbanism policies (see Rapoport 2015a) and the fact that globally mobile knowledge is often seen as expert knowledge. Rapoport (2015b) explains that when it comes to sustainable urbanism there is a perception that local expertise is lacking and that a global approach is needed. To the contrary, our book highlights that external factors – such as policy mobility from one international context to another – have little impact on the emergence and development of planning innovations for urban sustainability. As shown in the chapter on India (Chapter 11), the implementation of the Smart City concept – which is a highly mobile economic development model – was successful thanks to the inputs of the local stakeholders. This does not mean that planning innovations can only be explained by recognising the local factors that play a role in their emergence and evolution. Some of the planning innovations analysed in this book (Chapters 9, 11 and 13 specifically) also rely on funding and actors' networks operating at the national scale. Nevertheless, overall the impact of policy circulation is minimal. Montero (2011) in his review of 'mobile urbanism' indicated that urban sustainability policies were absent from the book's analysis. Our findings suggest that sustainability policies must integrate a level of complexity that makes the transfer of one policy model from one context to another less relevant.

This book, in a similar vein to the approach developed by Tenemos et al. (2018), illustrates future avenues for research on sustainability policies. It highlights the development of ad hoc actors' networks and financing innovations within the sustainability transition agenda rather than emphasising the influence of neoliberal policies, which is too simplistic as an analytical framework when it comes to understanding the factors influencing the development of innovations for urban sustainability.

References

Campbell, S. (1996) "Green cities, growing cities, just cities? Urban planning and the contradictions of sustainable development", *Journal of the American Planning Association* 62(3), 296–312.

Dixon, T. (2006) "Integrating sustainability into brownfield regeneration: Rhetoric or reality? An analysis of the UK development industry", *Journal of Property Research* 23(3), 237–267.

Loorbach, D. and Rotmans, J. (2010) "The practice of transition management: Examples and lessons from four distinct cases", *Future*, 42, 237–246.

McCann, E. and Ward, K. (eds) (2011) *Mobile Urbanism: Cities and Policymaking in the Global Age*. Minneapolis: University of Minnesota Press.

Montero, S. (2011) "Book review. Mobile urbanism: Cities and policymaking in the global age", *Berkley Planning Journal* 24(1), 168–172.

Rapoport, E. (2015a) "Sustainable urbanism in the age of Photoshop: Images, experiences and the role of learning through inhabiting the international travels of a planning model", *Global Networks* 15(3), 307–324.

Rapoport, E. (2015b) "Globalising sustainable urbanism: The role of international master-planners", *Area* 47 (2), 110–115.

Rutherford, J. and Coutard, O. (2014) "Urban energy transitions: Places, processes and politics of socio-technical change", *Urban Studies* 51(7), 1353–1377.

Tenemos, C., Baker, T. and Cook, I. R. (2018) "Inside mobile urbanism: Cities and policy mobilities". In Schwanen, T. (ed.) *Handbook of Urban Geography*. Cheltenham: Edward Elgar.

INDEX

For Product Safety Concerns and Information please contact our EU
representative GPSR@taylorandfrancis.com
Taylor & Francis Verlag GmbH, Kaufingerstraße 24, 80331 München, Germany

www.ingramcontent.com/pod-product-compliance
Lightning Source LLC
Chambersburg PA
CBHW070409270326

41926CB00014B/2761